Couplings and Shaft Alignment

Couplings and Shaft Alignment

by

Michael Neale
Paul Needham
and
Roger Horrell

**Professional
Engineering
Publishing**

Professional Engineering Publishing Limited
Bury St Edmunds and London, UK

First published 1991
Revised and reprinted 1998

ISBN 978-1-86058-170-0

British Library Cataloguing in Publication Data
Neale, Michael
  Couplings and shaft alignment.
  1. Mechanical components. Flexible couplings.
  I. Title   II. Needham, Paul   III. Horrell, Roger
  621.825

Contents

Preface

The need for a book such as this was postulated by the Institution of Mechanical Engineers. A literature search on couplings and shaft alignment revealed, in particular, the *Proceedings of the International Conference on Flexible Couplings* held in 1977, and a multitude of articles in various journals worldwide. The opinions of machine specialists in the oil, gas, petrochemical, and electricity generation industries were solicited, and there was general agreement that coupling users would benefit from a publication which brought together the scattered knowledge on couplings, and the inextricably-linked subject of alignment, into a single publication convenient for reference.

This book is the work of three authors, each having particular experience complementing that of the others. Michael Neale, of Neale Consulting Engineers Ltd, 43 Downing Street, Farnham, Surrey organized the original 1977 conference. He is perhaps best known as the editor of the *Tribology Handbook,* and for the Consultancy which bears his name and which specialises in the solution of problems with machinery. In this publication he contributed the chapters on the torsionally rigid couplings.

The chapters on elastomeric element couplings and torsional dynamic response are those of Paul Needham, of the Holset Engineering Company Ltd, 131 Parkinson Lane, Halifax. Holset's special expertise is in the application of torsionally soft couplings used, in particular, to control torsional vibration.

The chapters on shaft alignment are the work of W. Roger Horrell, of Romoil Ltd, 46 Redwing Lane, Stockton, Cleveland, who, through his family-owned company, provides machines expertise in design, installation, maintenance and problem-solving, to clients in the continuous process industries.

Thanks are due to many but, in particular, to the late Alan Axford who, until his untimely death, played a leading role in the opening chapters and in getting the book under way. Within Imperial Chemical Industries PLC, he was the foremost authority on machine design; he was diligent in recording his experience for others, and gave the authors encouragement and direction.

A note from readers drawing attention to the undetected errors in the text would be appreciated by the authors, as would other comment. The subject of couplings and shaft alignment is still evolving, and a free exchange of information and experience can do much to ensure long life and trouble-free running of industry's machines. Your response is invited.

Foreword

This book deals with flexible couplings which transmit power between the shafts of important rotary machines, and with shaft alignment. It does not cover couplings for the very largest machines such as turbo-generators for major power stations, nor couplings for small general-purpose machines, nor the special couplings used in mechanical handling and material processing equipment where impulsive loading is encountered.

The clear objective has been to provide positive practical guidance to those engineers involved in the design, specification, selection of equipment, and operation and maintenance of machine systems involving shaft couplings for the transmission of power, and the inter-related problems of shaft alignment. At the conceptual stage of process plant design, many alternative possible schemes are investigated. When finally a short list of feasible plant schemes (or individual machine train arrangements) are compared, their capital cost can be estimated with reasonable precision, but reliability is only quantifiable with difficulty. A systematic approach to the selection and specification of shaft couplings, and a rational approach to shaft alignment, are therefore essential to achieve the desired level of reliability. In the petrochemical and oil industries, reliability is sought through the application of specifications based on API (American Petroleum Institute) Standards. Large companies often employ specialist machines engineers who codify the collective experience of their company's plants into in-house specifications. In contrast, some buyers prefer to place themselves entirely in the hands of the principal machine vendor, notwithstanding the legal maxim of *caveat emptor* (let the buyer beware), and despite the fact that the limit on claims against equipment suppliers is usually only a fraction of the potential loss due to enforced plant shutdown. Thus, reliability remains the important issue.

This book is not intended *primarily* for coupling manufacturers, who face problems of detail design and production, but they may find it helpful to understand the approach of the process plant designer. Although some manufacturers may consider the guidelines unduly restrictive, there is no intention to inhibit development. Indeed, because of development and improvement, any guide based on service experience needs a rolling update by users, as more information becomes available.

Devices allied to shaft couplings, but dealt with in other publications, are variable speed couplings, freewheel or override couplings, clutches, and crankshaft dampers for reciprocating machines.

CHAPTER 1

An Approach to Coupling Selection

1.1 PHILOSOPHY

The strategy is simply to make choices that are favourable to obtaining the required intrinsic reliability. Thus we are concerned with rules governing design decisions, both for the devices and for the systems in which they are used.

Now, the procedures for making rational initial choices are nearly the converse of detail engineering design procedures. For example, conventional design techniques for components such as wheels, suspensions, gearboxes, permit their assembly into a self-consistent design for a railway locomotive which will have a certain performance. The contrasting strategy is to draw up a desired railway timetable and deduce the requirements for the rolling stock, adjusting both to meet bounding limits on performance for reliable operation.

For this latter strategy to work, there must exist generalised rules derived from proper analysis of past successes and failures. These rules are usually manifest as empirical limits for relevant parameters.

1.2 DEPTH OF ASSESSMENT

A general-purpose machine direct-coupled to an electric motor rated at less than 200 kW and supported by an installed standby unit is unlikely to repay the effort of particular assessment by a plant designer. Many organisations keep lists of preferred vendors selected on the basis of their providing a satisfactory product and meeting the criteria of cost and delivery time. No more need be done for such machines.

On the other hand, an un-spared major machine train for a single-stream process plant, with transmitted powers exceeding 2 MW, comprising several coupled units, clearly merits intensive assessment.

In between these two extremes cases, the engineer can only rely on his judgement. A gauge of importance, from the standpoint of the needed intrinsic reliability of the machine train, can be helpful.

1.3 ASPECTS OF RELIABILITY

The reliability engineering discipline has developed techniques of considerable power. These mostly depend upon having a statistically significant

number of event histories or the classified data banks exemplified by the
Euro-data system.

A link between failure rate and failure mode appears to be more depend-
able than a link between failure rate and a physical component of a
mechanical device. Accordingly, a complex device is best modelled as an
aggregate of failure modes. Design assessment strategy is, therefore, to
identify parameters that govern failure modes and find limits to such
parameters.

For example, from examinations of failed couplings of the gear type,
Boylan suggested that the tooth sliding speed was the crucial parameter
governing surface damage by worm-tracking.

Limits on such parameters are not to be construed as precise values, but
as representative values for a distribution, so that the risk of failure rapidly
increases as the limiting value is approached and exceeded.

1.4 PROCEDURE

Design is invariably based on iteration. The first step is to make a preliminary choice of the coupling type, or at least to rank the common types in
order of priority of attention. The chosen type can then be reviewed against
parametric criteria. This may show that the parameter closely approaches
or crosses some bounding value, leading to:

— a modified scheme;
— consideration of another type of coupling;
— investigation of the particular coupling offered by a vendor as proven
  for the proposed duty;
— development of a coupling to meet the duty.

1.5 INFLUENCE OF TIME PROGRAMMES

For large capital projects, the programme often demands firm decisions
surprisingly early in the design phase of the plant. The period for equipment optimisation through iteration is curtailed to permit serious design
work on plant layout and civil work.

There is, therefore, considerable pressure on the plant design engineer
to anticipate machine vendors' offers by specifying the preferred train
arrangement and preferred coupling types in the formal enquiry issued to
prospective vendors. This practice at least puts the vendors on an equal
footing when submitting offers.

CHAPTER 2

Preliminary Choice of Coupling Type

The process of making this choice is illustrated in Fig. 2.1. Notes on the various steps are given below.

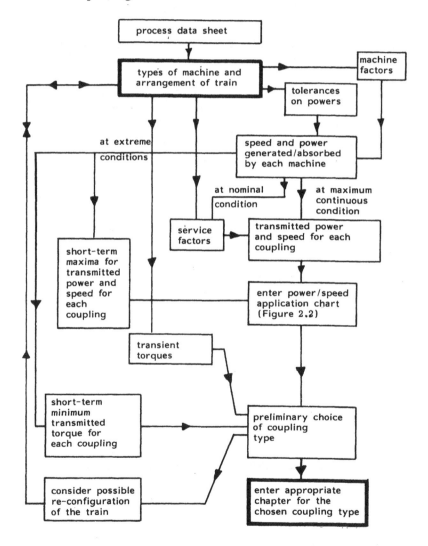

Fig. 2.1 Flowchart for the preliminary choice of coupling type

2.1 LONG TERM STEADY-STATE RUNNING CONDITION

We need paired values of power and speed, when the machine train runs at any process condition, within the expected range of continuous plant operation.

The process data sheets should provide a series of self-consistent sets of conditions. Using this basic data the appropriate machines are chosen and their powers and speeds estimated. The matching of machine type to duty is not part of this publication but the choice of couplings is a contributory factor in deciding the machine-train arrangement. It is necessary to consider the likely tolerances on these powers only when the transmitted power is a small residual from a driver and another driven machine. Note that machines purchased against API Standards are normally acceptable if their powers are within 4 per cent of guaranteed values.

The type of machine influences the coupling rating. For making the preliminary choice of coupling, apply the service factors given in Table 2.1. Of course, a full analysis should be made during the engineering phase of the machine train procurement, when this influence on rating, and the margins added at each stage of design, can be evaluated.

The process data is often limited to a single set of nominal conditions. Experience suggests that exigencies of plant production needs, together with the characteristics of the types of machine comprising the train, subject couplings to higher duties than those obtained from such nominal conditions. Traditionally, this difficulty has been met by applying these empirical service factors.

The individual machine powers and speeds give the corresponding transmitted power and speed for each coupling for the particular arrangement of the machine train under consideration.

In most cases the important parameter is the power at maximum continuous speed. Sometimes the power may reach a maximum value at a lower speed, and this data pair may prove more important when entering the power speed chart (Fig. 2.2).

2.2 SHORT-TERM RUNNING CONDITIONS

The process dictates the special conditions over brief periods, usually less than 48 hours, for plant start-up, shut-down, or emergency operation. This case is often the most important for trains with dual drives (e.g., having both an electric motor and a turbine) when the torque may reverse or become nearly zero.

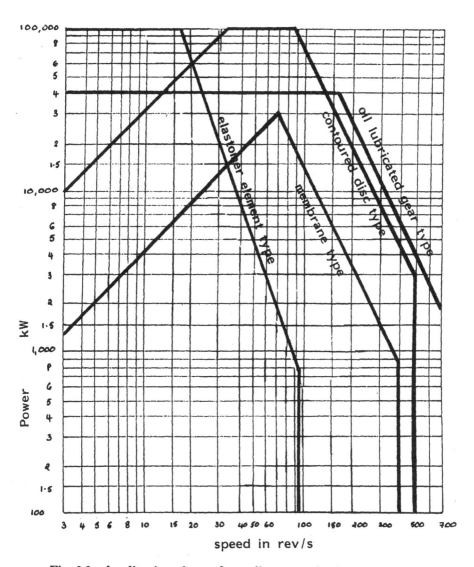

Fig. 2.2 Application chart of coupling types by factored power and speed

Table 2.1   Service factors
(From API 613, Special purpose gear units for refinery services)

Driven Equipment	Prime Mover		
	Motor	Turbine	Internal Combustion Engine
Blowers			
Centrifugal	1.4	1.6	1.7
Compressors			
Centrifugal	1.4	1.6	1.7
Axial	1.4	1.6	1.7
Rotary lobe (radial, axial screw, and so forth)	1.7	1.7	2.0
Reciprocating	2.0	2.0	2.3
Fans			
Centrifugal	1.4	1.6	1.7
Forced draft	1.4	1.6	1.7
Induced draft	1.7	2.0	2.2
Generators and exciters			
Base load or continuous	1.1	1.1	1.3
- Peak duty cycle	1.3	1.3	1.7
Pumps			
Centrifugal (All service, except as listed below)	1.3	1.5	1.7
Centrifugal — Boiler feed	1.7	2.0	—
Centrifugal — Hot oil	1.7	2.0	—
High-speed centrifugal (over 60 r/s)	1.7	2.0	—
Centrifugal — Water supply	1.5	1.7	2.0
Rotary — axial flow — all types	1.5	1.5	1.8
Rotary — Gear	1.5	1.5	1.8
Reciprocating	2.0	2.0	2.3

2.3 TRANSIENT TORQUES

Significant transient torques usually arise from fault conditions. They invariably occur when electric generators or motors are part of the train. When induction motors are started direct-on-line the standstill transient torque is of the order of 200 per cent of the motor full-load torque. The peak torque occurs during acceleration, commonly at a speed close to the normal operating speed and typically within the range 2.0–2.8 times the rated full-load torque. Even when special starting arrangements are used to reduce the starting current drawn from the supply, there may be transients when the associated switchgear operates. For example, in the Korndorfer system, three consecutive modes of motor operation are created by switching:

(1)   auto-transformer start (for supply at reduced voltage);
(2)   inductive impedance in supply line;
(3)   direct-on-line.

For the preliminary choice of coupling type these transient torques created by starting induction motors can be ignored when an elastomer element coupling is chosen. Not so for synchronous machines; these can produce transient torques of the order of six times full load torque in the event of mal-synchronisation of a generator, or 'pull-in' of a synchronous motor run-up as an induction motor. Rigid or quill couplings are then the first choice.

This impulsive transient torque should be carefully distinguished from the fluctuating torque produced over the whole of the run-up period when salient-pole motors are started by induction. Such a fluctuating torque constitutes an energetic excitation of the torsional natural frequencies of the train. High impulsive torques may occur on mechanical 'fault' conditions, especially in trains which have coupled flywheels to increase the moment of inertia.

It may be more economic to ignore large transient torques when rating the coupling and to install a torque limiter. This implies that the mean time between events requiring production downtime to renew the limiter is sufficiently long to be accepted. Such decisions involve operability studies, hazard analysis, and allied techniques.

2.4 CYCLIC TORQUE

Pumps and compressors of the positive displacement type invariably generate a cyclic torque fluctuation. For reciprocating machines the degree of fluctuation is reduced by providing a flywheel. For large machines rated at more than 1 MW the conventional arrangement uses a rigid coupling to a multipole synchronous motor which supplies the necessary flywheel effect.

Machines with lobed rotors, and small reciprocating compressors, are commonly connected to their drivers through torsionally soft couplings, especially when the machine is driven through a gearbox. The factors given in Table 2.1 include an allowance for the cyclic torque characteristic.

Axial and centrifugal compressors generate a cyclic torque fluctuation when operated in surge. No account is taken of this because the machine protection system prevents the operating point approaching the surge zone. Special consideration is needed for installations where surge protection systems are not provided.

2.5 PRELIMINARY CHOICE

The parameters influencing this choice are summarised in Table 2.2, together with corresponding ranking of suitable coupling types. Very high powers or very high speeds usually lead to the choice of a solid coupling, which may be rigid, or embody a quill shaft.

One view of the flexible coupling is that it is a practical device used to avoid the problems associated with a rigid coupling. The permissible degree of misalignment is best considered after making the preliminary choice of coupling type. The limits of misalignment can then be related to the constructional specification of the coupled machines and their baseplate or foundation. It is seldom necessary to re-select the coupling type against misalignment criteria unless some uncommon constraint is applied, such as operation following displacements due to earthquake phenomena.

Table 2.2   Parameters influencing preliminary choice of coupling type

Code	Parameter	Condition	Choice of coupling ranked by type
1	Factored* power at maximum continuous speed	Long term steady-state operation	Take from Figure 2.2
2	Speed at maximum factored* power		
3	Minimum torque	The transmitted torque closely approaches or passes through zero during steady-state operation	1 Membrane or contoured disc or quill 2 Elastomer-element
4	Transient torque	Coupling between parallel-shaft gearbox and synchronous motor or generator	1 Quill or Rigid 2 Elastomer-element
		Coupling between an epicyclic gearbox and an electric induction motor or generator	1 Elastomer-element 2 Quill 3 Gear
		Transient torque exceeds twice the value derived from code 2	Take from Fig. 2.2 or consider torque limiting devices with choices as codes 1 or 2
5	Paired values of power and speed for short-term special operating conditions	When any pair exceeds values from codes 1 or 2	Take from Figure 2.2.
6	Cyclic torque variation, as in reciprocating machinery	Long-term steady-state operation	1 Rigid 2 Elastomer-element or quill

* using factors from Table 2.1.

2.6 EXAMPLE

Reciprocating compressor driven through a gearbox by a steam turbine.

Description

This train comprises:

(a) a two-stage compressor having five cylinders arranged vertically in line;
(b) a double-reduction parallel-shaft gearbox using single-helical gears;
(c) a non-condensing pass-out turbine.

The expected long-term maximum continuous running conditions are:

	Compressor	*Turbine*
Power (kW)	4260	4370
Speed (r/s)	3.83	60

Choice of coupling type

Ordinarily, such reciprocating compressors are rigidly coupled to the rotors of synchronous electric motors. The use of a turbine driving through a separate gearbox leads to:

(a) relative displacements sufficiently large to cause unacceptable mis-alignment for a rigid coupling, making laterally flexible couplings the first choice;
(b) the need for continuously unidirectional torque transmission to avoid gear tooth chatter, making the first choice a torsionally resilient coupling, in conjunction with an appropriate flywheel on the compressor crankshaft.

From these considerations the preliminary choice of each coupling type is:

(a) an elastomer-element or a quillshaft coupling between the gearbox and the compressor flywheel;
(b) a spacer gear coupling between the gearbox and the steam turbine.

The final choice is the quillshaft type for both couplings. Both the user and the selected manufacturer have satisfactory previous experience with this type in this service, at comparable powers and speeds. Space is available to accommodate the quill lengths ($\sim$ 1m for the turbine coupling and $\sim$ 2m for the compressor coupling).

Torque measurement

During normal running, the steam condition from the turbine exit is slightly wet. The precision to which the turbine power can be inferred from thermal measurements is considered insufficient. Accordingly, provision is made for direct torque measurement by strain gauges applied to the quillshaft. These also permit direct measurement of transient vibratory torques.

Torque overload shearbolt

Both the quillshaft shear stress and the gear tooth loading stresses are proportional to torque. When the system was engineered, the instantaneous torque variation was limited to ± 30 per cent of the rated mean driving torque, but higher values can arise due to:

(a)   inadvertent continual operation inside the forbidden compressor speed band of 1–2.83 r/s, when resonance with natural torsional frequencies can occur (Note that on start-up the train accelerates through this speed band in a time sufficiently brief to avoid resonance effects.);

(b)   accidental seizure of the driver, when the considerable inertia of the flywheel (50 000 kg.m²) would cause overload of the gear teeth.

Accordingly shearbolts on the compressor quillshaft attachment are specified, to limit the maximum attainable torque to 300 percent of the rated mean driving torque.

When the shearbolts function, the drive is disconnected. This does not introduce a consequential hazard due to overspeed because:

(a)   reciprocating compressors of this type cannot operate as motors;

(b)   the turbine protection system includes independent overspeed trips.

Renewing shearbolts is a preventive maintenance action requiring shutdown of the compressor set, but it can be done concurrently with other planned outages. Downtime due to an intrinsic failure of shearbolts is ignored because such an event is considered a rare occurrence.

CHAPTER 3

Gear Couplings

3.1 DESCRIPTION

A typical gear coupling is shown in Fig. 3.1. The two hubs, which fit on to the shafts of the two coupled machines, incorporate gear teeth that mesh with an intermediate shaft or sleeve. Any kind of lateral or angular misalignment between the machines can be taken up by a combination of angular misalignments at each gear mesh. These angular misalignments are limited by the operating conditions at the gear meshes, and these limitations, in conjunction with the spacer length, determine the misalignment which the coupling allows between the coupled machines. Quite large axial movements can be permitted.

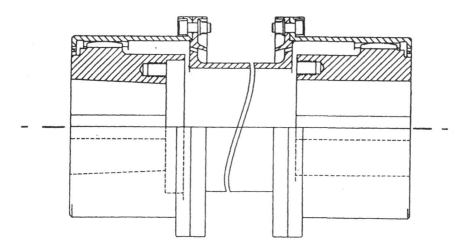

Fig. 3.1   A typical gear coupling for use at high power and speed

Involute gear tooth profiles are used almost exclusively, principally because they can be produced on standard gear manufacturing equipment. They also have the advantage of providing uniform velocity transmission. Pressure angles of 20–25 degrees, and 75–70 per cent full depth teeth are generally used for high torque, high speed couplings.

With involute tooth profiles, the two gears of a mesh have teeth which can be viewed as having been generated from a common base circle. When the two gears are in aligned concentric contact, the radii of curvature of the contacting tooth surfaces are identical and the teeth make full area contact. However, for couplings which are expected to operate with substantial misalignment, the external or male teeth are usually barrelled to give them curvature in an axial direction and avoid hard contact at the ends of the teeth. With barrelled teeth the contact length is shorter than with aligned straight teeth, but remains more constant with changes in alignment.

Note

The worst loading conditions for
the teeth occur when the tooth active
flanks are exactly at right angles to
the plane of misalignment. ie at
90+∅° and 270+∅° where ∅ is the
tooth pressure angle.

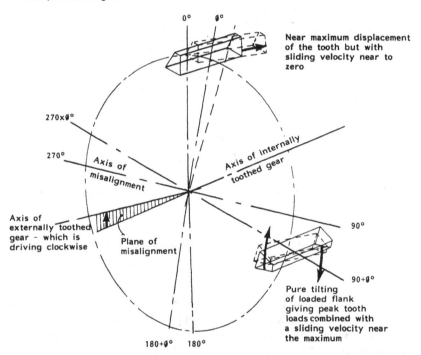

Fig 3.2   Tooth movements due to misalignment of a gear coupling mesh

Figure 3.2 shows the general nature of the contact pattern around the teeth of a misaligned gear coupling. Teeth which are passing through a position where their active surfaces are parallel to the plane of misalignment, near to the 0 and 180 degree positions, have potential contact over the maximum tooth area and their relative displacement is of a pure sliding form. Teeth at 90 degrees to these first teeth, and which have their active surfaces at right angles to the plane of misalignment will not have any relative axial displacement, but will be tilted relative to each other, so that the contact loads will tend to be concentrated towards the ends of the teeth. Teeth at intermediate positions will have various combinations of axial displacement and relative tilt.

The teeth in a gear coupling mesh are not uniformly loaded because the tilted tooth contact, concentrated at around the 90 and 270 degree positions, produces relative circumferential displacement between the two gear rings. The teeth with the greatest tilt carry the greatest load and in fact the teeth at around the 0 and 180 degree positions may well be out of contact with each other.

The torque capacity and allowable speed of gear couplings is dictated primarily by the operating conditions which the teeth have to endure when passing through the regions around the 90 and 270 degree positions. In these regions the maximum tooth contact stress occurs, due to the combined effects of uneven circumferential load sharing and tooth edge loading. The relative sliding velocity between the teeth is also at a maximum at around these positions, because they lie half way between the maximum 'static' tooth displacements at around the 0 and 180 degree positions.

Checking the suitability of any gear coupling for a particular installation will, therefore, involve checks on tooth contact stresses and sliding speeds, in relation to the material specifications for the teeth. Also since the peak loads and sliding movements occur simultaneously, lubrication of gear couplings is essential and requires careful attention. These various aspects are covered in the following parts of this chapter.

3.2 SELECTION

The factors which determine the selection of an appropriate gear coupling for a particular application are the maximum rotational speed and the maximum torque or power that has to be transmitted, together with the required misalignment capacity. The speed and torque influence the allowable maximum and minimum diameter, respectively, while the misalignment capacity can be adjusted by the selection of an appropriate spacer length.

3.2.1 Power and speed requirements

The maximum torque which can be transmitted by a gear coupling is determined by the stresses on the teeth. Couplings with a larger pitch circle diameter (PCD) can therefore transmit more torque, and an indication of the range of performance obtainable is given in Fig. 3.3. A coupling of a given diameter is also limited in the maximum speed at which it can operate, by centrifugal stresses, and this determines the right hand limit of the performance envelope in Fig. 3.3. In very high power couplings the stress patterns which limit the ultimate performance become rather more complex, but in practice, certainly in the process industry, a maximum power of 40 MW is usually taken as a conservative upper limit for gear couplings for convenience and reliability. Above this power, rigidly coupled shafts are more likely to be used.

When selecting an appropriate coupling size for a given duty, Fig. 3.3 gives an indication of the minimum pitch circle diameter of the teeth which will transmit the power at the operating speed. This will also give the lightest coupling and, therefore, the one which is the least likely to cause vibration problems or lateral critical speed problems on the coupled machines. Smaller diameter couplings are also easier to lubricate because the centrifugal forces on the lubricant are less for any operating speed. However, if considerable axial movement is envisaged, the resulting frictionally derived axial loads on the thrust bearings of the coupled machines will be less, if a larger PCD coupling is used. This is because, for a given torque, the tooth loads and, therefore, the tooth axial friction displacement forces will be less with a larger diameter. In some cases, therefore, it may be desirable to pick a coupling that is a little larger than the minimum allowable from tooth stress limits. The maximum size of coupling which can be used is given in Fig. 3.3 by the point on the centrifugal stress limit line corresponding to the maximum operating speed. In the example shown on the figure a coupling to transmit 2MW at 200 r/s would need a tooth PCD of about 100mm but if desired a coupling with a PCD of up to 200mm could be used.

3.2.2 Allowable misalignment and axial displacement

The physical limits which determine the allowable angular misalignment at a gear coupling mesh are shown in Fig. 3.4. The clearance between the teeth provides a further limit, which is determined by the coupling designer. It will usually be kept as tight as possible, to reduce any run out and imbalance, and will generally be set at a maximum value which is close to those determined by the physical limits.

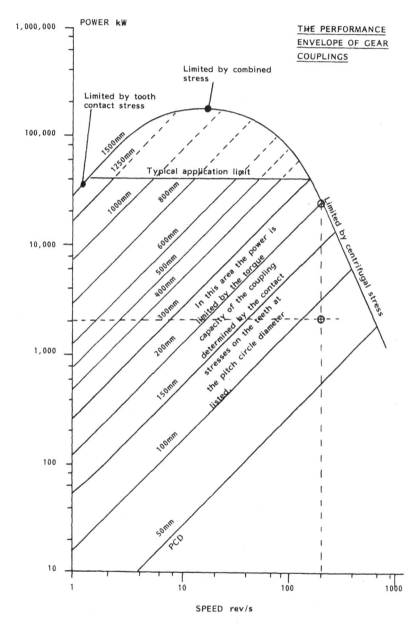

Fig 3.3   The performance envelope of gear couplings

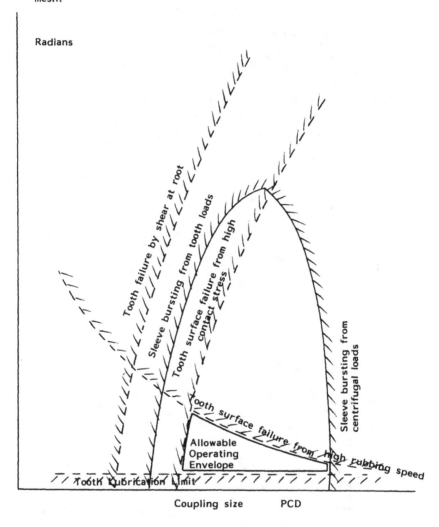

Allowable angular
misalignment across
a gear coupling tooth
mesh.

Radians

Tooth failure by shear at root

Sleeve bursting from tooth loads

Tooth surface failure from high contact stress

Sleeve bursting from centrifugal loads

Tooth surface failure from high rubbing speed

Allowable
Operating
Envelope

Tooth Lubrication Limit

Coupling size   PCD

**Fig 3.4   The physical limits which determine the allowable
misalignment and coupling size to carry a specified power and
speed**

Figure 3.4 shows that the main physical limit on the angular misalignment at a gear coupling mesh is a critical value of the mean rubbing speed at the tooth contacts. It has been found by experience that the maximum value of this rubbing speed should not exceed 0.12 m/s. If this value is exceeded, the tooth surfaces tend to wear excessively and usually produce a form of surface damage, commonly called worm tracking, because it is similar in appearance to the marks left by woodworm on antique furniture. Actual values of angular misalignment which correspond to this limit are shown in Fig. 3.5 for couplings of various sizes operating over a range of speeds.

The other potential physical limit on the maximum misalignment is the tooth contact stress. This is increased by misalignment, because of the combined effect that misalignment has on increased edge loading, unless the teeth are barrelled, and on uneven circumferential load sharing between the teeth.

The overall effect on the tooth contact stress is, however, small, relative to that which already arises from aligned torque transmission, as indicated by the near vertical nature of the tooth surface contact stress line in Fig. 3.4. In practice, therefore, once a coupling has been selected to be of sufficient size to carry the required power and speed, the *maximum* allowable misalignment can be assumed to be determined entirely by the tooth rubbing speed limits given in Fig. 3.5.

There is, however, also a *minimum* recommended value of angular misalignment at each coupling gear mesh, to prevent the teeth remaining in constant static contact with the exclusion of lubricant. A minimum angular misalignment of the order of 0.00075 radians at each mesh helps to give some relative movement between the teeth and sufficient load reduction at around the 0 and 180 degree positions, as in Fig. 3.2, to assist in the replenishment of the lubrication of the tooth surfaces.

The axial displacement which a gear coupling allows is determined by the detailed design and, in particular, by the excess length of the internal teeth. Some form of positive limitation to the axial movement of the spacer is required to ensure that the coupling remains fully in mesh. In some installations the coupling is used to locate the shafts axially with respect to each other. With such limited end float couplings the shaft ends may be provided with hardened spherical-ended buttons to provide additional location, with minimum risk of fretting.

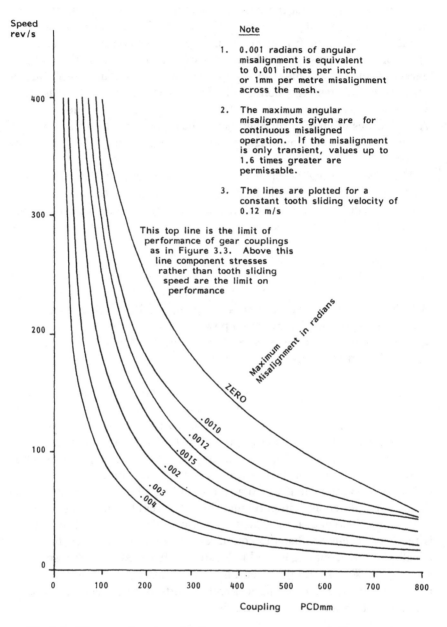

Fig 3.5 The maximum misalignment at a gear coupling mesh to avoid excessive tooth wear

3.3 CHECKING A COUPLING FOR SUITABILITY

After a particular coupling has been selected and its dimensions are known, it can be checked for suitability in terms of tooth stress. The contact stress on the teeth of a gear coupling can be calculated from

$$S_c = \frac{2TF_sK_m}{D_TZd_tw_T} \times 10^3 \text{ MN/m}^2$$

where

T = torque (Nm)
F_s = service factor as in Table 2.1
K_m = load distribution factor (see Table 3.1)
D_T = pitch circle diameter (mm)
Z = number of teeth
d_t = tooth depth (mm)

Note that, for any particular coupling, D, Z, and d_t are interrelated.

Table 3.1   Values of K_m

Angular misalignment	Load distribution factor, K_m, with face width, w_T equal to:			
i (radians)	12.5mm	25mm	50mm	100mm
0.001	1	1	1	1.5
0.002	1	1	1.5	2.0
0.004	1	1.5	2.0	2.5
0.008	1.5	2.0	2.5	3.0

i = angular misalignment (radians)
w_T = full width of the teeth (mm)

Table 3.2   Maximum allowable values of tooth contact stress S_c'

Material	Hardness	Allowable stress in MN/m²
	(H_v)	S_c'
Steel	160–200	5
Steel	230–260	7
Steel	300–350	11
Flame and induction hardened steel	500–590	14
Nitrided steel	600–650	17

3.4 FORCES AND MOMENTS GENERATED

3.4.1 Axial forces

When a coupling is transmitting torque, relative movement between the mating teeth is resisted by friction at the tooth contacts, and for smooth, lubricated steel surfaces the coefficient of friction in slow speed sliding at any individual tooth contact will be about 0.15. However, the apparent friction coefficient for the whole coupling due to externally imposed axial displacement – for example, due to thermal expansions of the connected machines – depends on the sum of all the tooth friction forces, and when there is significant misalignment at the tooth meshes, the friction forces on some of the teeth will be assisting the displacement. In these circumstances equivalent coefficients of friction as low as 0.05 can occur. On the other hand, if the coupling is operating in good alignment and the required axial movement is relatively sudden, then the full friction coefficient of 0.15 may well be applicable.

The value of the coefficient of friction obtained is important because it is a critical factor in specifying the required load capacity of the thrust bearings of the coupled machines. If the coupling teeth have become worn in service, a coefficient of friction as high as 0.3 could arise. Such high coefficients of friction are, however, usually only associated with inadequate lubrication and maintenance. It is believed that it could also occur with a coupling operating in perfect alignment so that oil is excluded from the operating surfaces and dry friction is occurring. It should be noted that the angular alignment does not have to be perfect for the teeth to lock, since the shafts and spacer are capable of bending slightly to cater for slight misalignment. This emphasises the need for rigid coupling designs, and also implies that one possible solution is to avoid near-perfect alignment.

If the possibility of near-perfect alignment cannot be avoided, three options are available. Firstly, where a very high level of reliability is essential, the thrust bearing must be designed for $\mu = 0.3$. If this makes the thrust bearing design too large and subject to excessive power loss, it may be possible to overcome the problem with a larger PCD coupling, although the extent to which this can be done may be limited by the increased weight, tooth sliding speed, and centrifugal stresses. A larger coupling reduces the net tooth force and, hence, the thrust transmitted with a given friction coefficient. If, even then, an adequate thrust bearing design cannot be produced, the second option is simply to accept the possibility of very infrequent thrust bearing overload and to design for a lower friction coefficient of, say, 0.15. The difficulty here is that the risk of failure cannot really be properly quantified, and if this option is unacceptable, the only remaining alternative is not to use a coupling of the gear type.

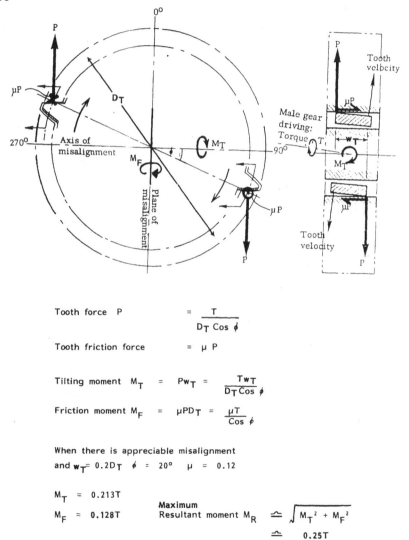

Tooth force $P \quad = \dfrac{T}{D_T \cos \phi}$

Tooth friction force $= \mu P$

Tilting moment $M_T = Pw_T = \dfrac{Tw_T}{D_T \cos \phi}$

Friction moment $M_F = \mu P D_T = \dfrac{\mu T}{\cos \phi}$

When there is appreciable misalignment

and $w_T = 0.2 D_T \quad \phi = 20° \quad \mu = 0.12$

$M_T = 0.213T$

$M_F = 0.128T$

Maximum

Resultant moment $M_R \; \cong \; \sqrt{M_T^2 + M_F^2}$

$\cong \; 0.25T$

Fig 3.6 The moments generated at a gear coupling mesh with appreciable misalignment. All forces shown are those on the female gear. Those on the male gear are equal and opposite.

3.4.2 Transverse forces

When machines are coupled together with gear couplings substantial radial forces can be applied to the journal bearings of the machines. These forces arise from transverse moments which are generated at the meshes of the couplings, due to the tilting and sliding action of the teeth as shown in Fig. 3.2. Figure 3.6 shows how a tilting moment arises at the mesh from the edge loading of the tilted teeth, and a friction moment from the relative sliding movement of the teeth. Figure 3.6 also gives the approximate magnitude of these two moments and the maximum resultant moment for a coupling with appreciable misalignment.

In practice, with high speed gear couplings, as used in most process industry applications, the actual permissible misalignment is severely limited. Consequently, when some deflection is allowed for, the loading does not usually reach the extreme ends of the teeth, and rather lower moments are generated. With a typical value of tooth width of 0.2 of the PCD, and a coefficient of friction at the teeth of 0.12, it is reasonable to assume maximum resultant moments given by

$M_R = 0.16\ T$ for straight tooth couplings

$M_R = 0.12\ T$ for barrelled tooth couplings

Where T is the driving torque

If there is no misalignment, these moments will be zero.

The moments generated at the tooth meshes give rise to different forces at the bearings depending on the pattern of misalignment between the coupled shafts. Figure 3.7 shows that with a C pattern of misalignment the shafts of the machines are subject to the resultant moments only, while, with a Z or parallel offset misalignment, additional lateral forces also arise and can generate substantial bearing loads.

With this worst, and unfortunately common, situation the machine bearings adjacent to the coupling can be subjected to loads from the coupling given by

$$F = \frac{M_R}{L} \left\{ 2 + \frac{(L + 2a)}{b} \right\}$$

Downwards on the higher machine.

Upwards on the lower machine

Both at an angle θ to the vertical as shown in Fig. 3.7.

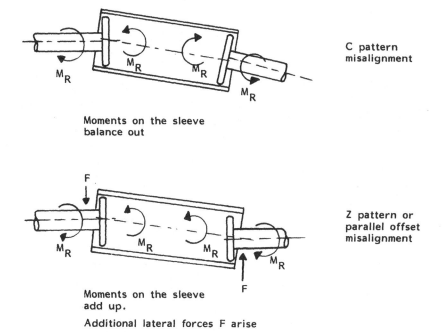

C pattern
misalignment

Moments on the sleeve
balance out

Z pattern or
parallel offset
misalignment

Moments on the sleeve
add up.

Additional lateral forces F arise

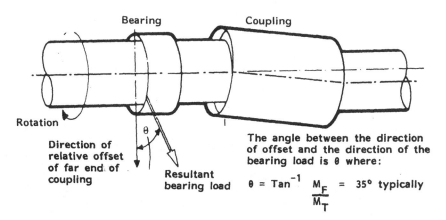

Bearing   Coupling

Rotation

Direction of
relative offset
of far end of
coupling

Resultant
bearing load

The angle between the direction
of offset and the direction of the
bearing load is θ where:

$$\theta = Tan^{-1} \frac{M_F}{M_T} = 35° \text{ typically}$$

**Fig 3.7   The effect of the misalignment pattern on the
magnitude and direction of the forces and the moments on the
shafts of machines connected with gear couplings.**

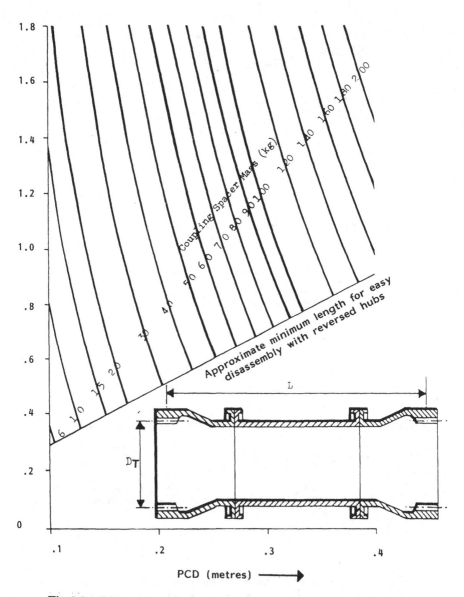

Fig 3.8   **Estimation of spacer mass for a high speed coupling design.**

where

M_R = resultant moment at the coupling mesh
L = length of coupling spacer
a = coupling mesh overhang from bearing
b = bearing span of the machine

and the corresponding load on the machine bearing remote from the coupling will be

$$\frac{M_R}{L}\frac{(L + 2a)}{b}$$ Upwards on the higher machine

Downwards on the lower machine

Both at an angle θ to the vertical as shown in Fig. 3.7.

In some machines these bearing loads generated by the coupling can be of the same order or greater than the loads arising in the machine itself, and can, therefore, be highly significant; the total bearing load may be increased to such an extent that the bearing is overloaded, or it may be decreased such that the shaft becomes unstable.

3.5 BALANCING AND CENTERING

Gear couplings need to be balanced to at least the same standard as the rotors of the machines to which they are coupled. It is, however, difficult to achieve good balance because the coupling cannot generally be balanced in the assembled condition since the spacer is a loose part and will only be properly centred when it is transmitting a torque.

The best that can normally be achieved is a balancing of the hubs, and of the spacer separately, using its outside diameter as a reference diameter on the balancing machine. The balance of the whole assembly in the torque-centred running condition then depends on the degree of concentricity between the various gear pitch circles and the balancing reference diameters, and the degree of freedom from cumulative pitch errors. Both must be very high for really good balance.

As with all couplings, when a three piece spacer is used, the location between the parts of the spacer must be accurate and repeatable, using either spigots or fitted bolts. Fitted bolts may be preferable for high speed applications because of the possible loss of spigot interference due to relative centrifugal growth of the parts.

The fact that the spacer will be subject to some degree of eccentricity also emphasises the need for low spacer mass. It is, therefore, important to use the smallest coupling size for the required duty, and the spacer must be in the form of a tube of high strength material with the lowest wall thickness consistent with reasonable resistance to handling damage and acceptable shear stress. A wall thickness of about 5mm is probably about the minimum acceptable for use in an industrial application. For some machines, considerable weight saving can be achieved by using a single piece spacer, but this greatly complicates assembly and disassembly, and is often impracticable.

For machines operating just below their lateral critical speed, lower spacer mass, together with low overhung length, is also important to minimize any reduction of the lateral critical speed of the shaft, which would bring it closer to the running speed. A reversed hub design with the teeth as close as possible to the machine casing, to reduce overhung length is therefore recommended, and hub mass should be minimized by use of high strength materials. Hubs fitted by oil injection will also be advantageous since avoidance of keys should allow smaller hub outside diameters to be used. Hubs which are integral with their shafts may be even better, provided that they do not interfere with the fitting of bearings, seals, and so on.

Because the spacer is located by its teeth under torque-generated forces it is important that the transmitted torque does not fall transiently to low values which might allow the spacer to move off centre. If this is likely to occur, direct centering on the tips of the teeth may be necessary. If tip centering is used, the tooth tips need to be accurately produced if interference between the tip centering and flank centering is to be avoided, and subtle design may be necessary to maintain effective tip centering at high speed because of the tendency for the female gear to grow relative to the male gear due to centrifugal strains. For these reasons, tip centering is probably best restricted to applications where there is a definite low-torque/high-speed running requirement, and to provide a quantitative guide it is suggested that tip centering need only be considered where

$$\frac{25T}{D_{T}MjN^{2}} \quad \text{is less than one}$$

T = Torque at speed, N (Nm) j = backlash at gear meshes (mm)
D_{T} = Pitch circle diameter (mm) N = rotational speed (r/s)
M = Spacer mass (kg)

For applications where the spacer mass is not known, an indication of the likely mass is given in Fig. 3.8.

Vibration can also arise if the coupling meshes become locked; at low misalignment the shafts ends then bend to accommodate the misalignment. The likelihood of this occurring can be reduced by the use of stiff shaft ends, minimum overhangs, and reversed hubs, in conjunction with the lightest possible spacer and good lubrication.

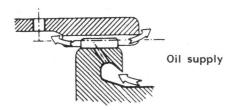

Oil supply

Continuously Lubricated – Damless Design

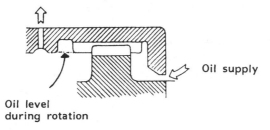

Oil supply

Oil level
during rotation

Continuously Lubricated – with Lubricant Dam

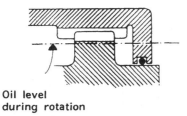

Oil level
during rotation

Grease or Oil Filled

Fig 3.9 Lubrication requirements for gear couplings.

3.6 LUBRICATION

Because the teeth of a gear coupling are subjected simultaneously to high contact loads and rubbing action, it is essential that they are lubricated. It is estimated that 75 per cent of gear coupling problems arise from improper or insufficient lubrication.

The various methods of tooth lubrication are shown in Fig. 3.9. For long periods of unattended reliable operation at high speeds, oil must be used rather than grease, and a continuous feed to each tooth and a damless coupling design are essential. Grease separates at relatively low speed, as indicated in Fig. 3.10, and is susceptible to hardening, which can prevent it flowing to the teeth. If grease *is* used the polyglycol greases tend to be better in terms of lubricant life and resistance to leakage past the seals. (The more solid components of the grease centrifuge to the outside, leaving the oil on the inside). Both oil- and grease-filled couplings are unsuited for long periods of operation because of seal unreliability and gradual loss of charge. A continuous flow is also necessary at high speed, both for cooling and, in conjunction with a damless coupling design, to prevent sludging problems which occur at high '*g*' due to centrifuging out of additives, impurities, and dirt in the oil.

For long periods of unattended operation, a suggested size–speed limit above which the use of damless coupling designs with a continuous lubricant feed is recommended, is indicated by the chain dotted line in Fig. 3.10. Guidance on the flow rates required with these designs is also given in Fig. 3.10. Damless coupling designs are liable to rapid deterioration in the event of a lubricant supply failure, and although this is probably of minor significance, since the oil feed will generally come from the bearing lubrication system, which will be provided with a low oil pressure trip, it does mean that careful design and installation of the piping to the coupling oil feed is essential. Also, with damless designs, lubrication may be impaired if a lot of wear is allowed to develop, due to the over-easy escape of oil past the teeth; this is a reason for changing couplings before major wear develops.

The grade of oil used has some effect on tooth wear, and it has been found that wear reduces as the viscosity is increased up to a viscosity grade of 680 or SAE EP140. Ideally the viscosity of the oil at the coupling operating temperature should not be less than 50 cp. In practice, however, the choice of oil for a continuous supply will normally be restricted to the limited range acceptable to adjacent parts of the system, since it will not generally be desirable to use an independent lubrication system for the

NOTE: Damless coupling designs, with
continuous lubricant feed to each tooth
must be used for applications above the
oil centrifuging limit and it is recommended
that they should also be used above the
chain dotted line where long periods of
unattended reliable operation are
required.
Flow rates indicated are for this type
of design.

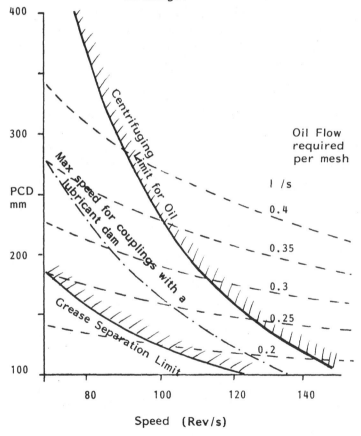

Fig 3.10   The lubrication requirements of gear couplings.

couplings. The need for a high viscosity does emphasise the need for a cooled oil supply, however. The supply also requires filtration to as fine a value as the oil system will allow, which will probably be about 10 microns nominal level. Attempts should not be made, however, to fit finer filters in the feed system to the coupling, because these will only block up by collecting the debris from the whole oil charge.

3.7 FAILURE MODES

The most common failure mode of gear couplings is excessive tooth wear, which often leads to indentation of the longer female teeth in the coupling sleeve, consequent lock up of the coupling against any axial movement, and thrust bearing failure. This most frequently arises from inadequate lubrication arrangements, which allow sludge to build up in the coupling teeth, or the loss of lubricant from oil filled couplings.

In highly stressed couplings tooth wear can also give rise to stress concentrations in the sleeve, which can result in fatigue failure of the sleeve itself in the vicinity of the teeth.

Failure is usually progressive and detectable by the monitoring of vibration levels. The sudden loss of drive without any warning is rare.

CHAPTER 4

Multiple Membrane Couplings

4.1 DESCRIPTION

Two types of multiple membrane coupling are shown in Fig. 4.1. The hubs which fit on to the shafts of the two coupled machines are connected to an intermediate spacer by flexible members. These members are made up from stacks of thin laminations so that they are flexible in bending and strong in tension and shear.

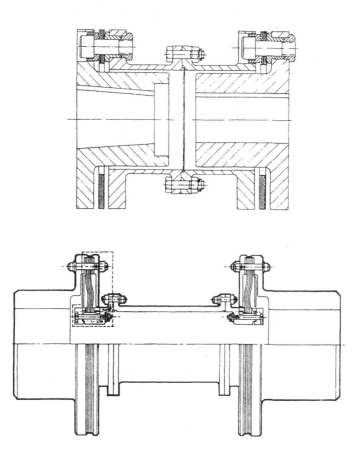

Fig 4.1 Typical multiple membrane flexible couplings.

The types of flexible members commonly used are shown in Fig. 4.2. The convoluted and radial-spoke type are connected to one shaft at the inner diameter and to the other shaft at the outer diameter. The driving torque is transmitted in shear across the diaphragm. The third type is in the form of a ring and the two shafts are connected to alternate holes in the ring with the torque being transmitted in tension along the portions of the ring forming links between the bolts. Axial displacement causes the ring to take up a circumferential wavy shape, when observed from the side, and angular misalignment creates a lack of symmetry in the waviness.

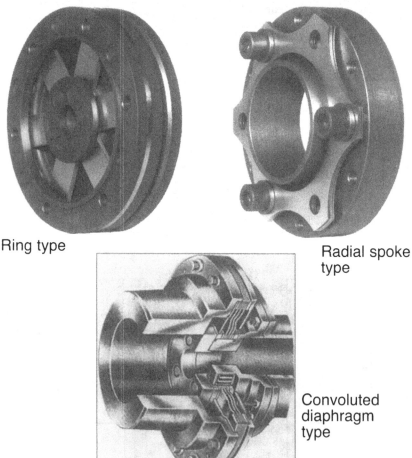

Ring type

Radial spoke
type

Convoluted
diaphragm
type

Fig 4.2   The main types of flexible membrane packs.

Multiple membrane couplings with ring type members tend to be the lightest and most compact because the circumferential tension link structure tends to be the most economical in the use of flexible material. With couplings of this type, if the flexible member were to fail, the fixing bolts will still transmit the drive. In the case of the radial spoked membranes, the radial arms would be the parts that would fail, and the inner part would then be free to rotate inside the outer. Such a coupling can, therefore, provide a form of ultimate torque limiting device in any application which may require it. For very high speed machinery, however, the extra mass of couplings fitted with radial membranes can be a disadvantage on machines operating just below a lateral critical speed, because the extra mass of the coupling at the end of the shaft overhang can bring the critical speed of the combined system closer to the running range.

Ring shaped membranes also allow a slightly greater misalignment capacity for a given coupling size and, other aspects being equal, these tend to be the preferred type for most applications.

The ability of the membranes in multiple membrane couplings to allow angular misalignment, combined with the use of an intermediate spacer shaft, means that any pattern of misalignment between two coupled machines can be taken up. The torque capacity is at a maximum when the machines are perfectly aligned. It is reduced slightly by axial displacement because this produces fatigue bending conditions in the links with flexing at a frequency corresponding to the shaft rotational speed.

The clamping conditions created by the bolts which attach the flexible members to the shaft components are particularly critical for satisfactory fatigue life, and the design of the coupling must be arranged so that these do not need to be undone in the field after controlled clamping in manufacture. Some additional demountable shaft coupling arrangement is, therefore, necessary to enable assembly and disassembly in service, without disturbing the clamping of the membranes.

It also needs to be remembered that the coupling spacer shaft represents a mass that is supported on two axial springs, and could, therefore, potentially be excited at its resonant frequency into an axial oscillation. It is generally considered advisable to keep this frequency above the running speed of the coupling, although, in practice, the risk of a resonant vibration is reduced because the axial spring stiffness of the membranes varies with their axial displacement. A high amplitude axial vibration is not, therefore, particularly likely.

Ring-type couplings with four bolts through the membrane, corresponding to two bolts per shaft, do not transmit a uniform motion when misaligned and are not recommended for most applications. For typical high speed rotating machines, as used in the process industry, it is recommended that multiple membrane couplings should have at least six bolts per membrane, corresponding to three bolts per shaft.

4.2 SELECTION

As an initial guide to coupling selection, a coupling size needs to be selected which has an adequate torque capacity. The minimum possible diameter is desirable because this will have the lowest mass and the smallest effect on machine lateral critical speeds and unbalance.

The misalignment capacity of the coupling also needs to be known so that it can be matched to that which the application requires. If misalignments towards the maximum allowable values are necessary, a slightly larger coupling may be needed compared with that which is suitable for the torque transmission alone.

4.2.1 Power and speed requirements

Figure 4.3 gives an approximate guide to the power and speed capacity of multiple membrane couplings of various sizes. These values are the maximum capacities with good alignment. The actual performance of individual couplings may vary a little from these values, according to detailed design features of the membranes used by the various manufacturers.

4.2.2 Allowable misalignment and axial displacement

The maximum allowable angular misalignment of a ring type multiple membrane coupling depends, among other things, on the number of bolts in the ring membrane and typical values are indicated in the table.

Number of bolts in the ring	Maximum angular misalignment (Radians)
6	0.012
8	0.009
10	0.006
12	0.004

These are maximum values, and the actual allowable values will be limited by fatigue stresses in the membranes corresponding to the combined angular and axial displacements at each membrane. The relative axial displacement between the two coupled machines will tend to be shared equally between the two membranes, but the angular displacement at each will depend on the combination of angular and lateral misalignment between the machines and the length of the spacer.

The form of the relationship between the allowable angular misalignment and axial displacement is shown in Fig. 4.4.

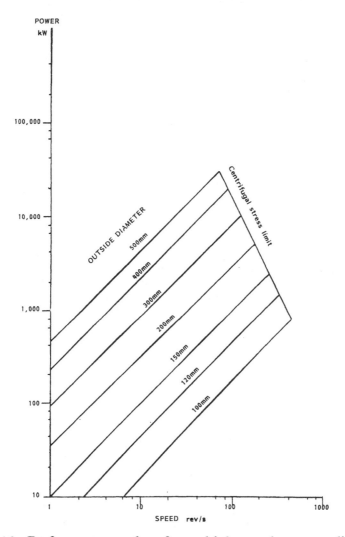

Fig 4.3 Performance envelope for multiple membrane couplings.

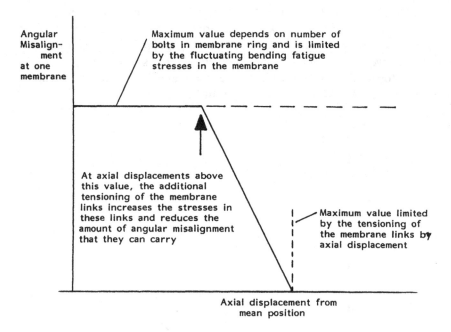

Fig 4.4 The factors which limit the allowable angular
misalignment and axial displacement of a multiple membrane ring.
(*Note.* This diagram shows the general form of the relationship
between allowable angular misalignment and displacement. The
precise nature of this relationship, according to the literature
of the various vendors, varies from a straight line to various
curves. The data for the particular coupling finally selected
should therefore be used.)

4.2.3 Coupling arrangement

The simplest coupling arrangement is as shown in Fig. 4.5(a), in which
there are two flanged hubs and a flanged spacer, with the flexible
membranes bolted to each component. On high speed machines, however,
it is desirable to have an additional pair of flange connections in the spacer
so that the coupling can be removed without having to undo the membrane clamping bolts. This is necessary because the tightening torques on
these bolts are critical and, also, if they are undone the standard of balance
of the coupling tends to be altered.

For particularly high speed systems, where critical speed may be a problem, a better arrangement is to use a reversed hub arrangement as shown in Fig. 4.5(b). This reduces the overhung masses and also ensures that all the parts are fully contained against any flailing action in the relatively unlikely event of membrane or bolt failure in service.

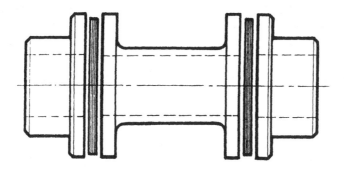

(a) Standard arrangement

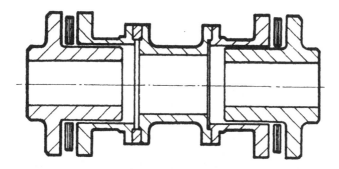

(b) Reversed hubs arrangement

Fig 4.5 Typical coupling arrangements.

4.3 DETAIL DESIGN CONSIDERATIONS

4.3.1 Coupling action

A typical membrane pack is illustrated in Fig. 4.7. It is assumed to be rotating anti-clockwise with the driving member in front of the paper connected to the A bolts, and the driven member behind the paper connected to the B bolts. Under torque, anti-clockwise, four of the eight links in this eight-bolt design will be in tension, and will tend to stretch (L_1 to L_4), whilst the other four (U_1 to U_4) will be compressed and, because they are made up of thin membranes, will tend to buckle and contribute little to transmitting the torque.

While the 'L' links are referred to as the loaded links, and the 'U' links as the unloaded links, in fact, the unloaded links may carry some tension if the coupling is sufficiently misaligned. When a coupling, which is carrying torque, is axially misaligned, the slack in the unloaded links is progressively absorbed until, with sufficient axial displacement, they too begin to be stretched. Up to this point the tension in the loaded links is largely unaffected by the axial displacement, but the maximum tensile stress in the individual membranes of the loaded links will be increased by bending. All membranes will carry the same stress pattern, since in these conditions of pure axial misalignment the ends of the links stay parallel.

4.3.2 Membrane stresses

Although the stresses due to bending of the individual membranes might be expected to be small, since the membranes are so thin, the effect of tension in the link is to concentrate the bending near the link ends immediately adjacent to the clamping washers, so that the bend radius can be quite small. The resulting maximum tensile bending stresses at this critical point in the membranes can consequently be several times the mean tensile stress, although there is scope for reducing these bending stresses by controlling the clamping washer chamfer radius or by using clamping washers with thin springy edges, as shown in Fig. 4.6.

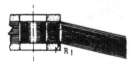

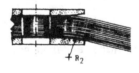
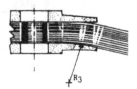

Square edge washers   Chamfered washers   Spring edge washers

Fig 4.6   Washer types.

In operation, the maximum stress in the free parts of the membranes of an angularly misaligned coupling occurs adjacent to the edge of the clamping washer at the surface of an outer membrane of a loaded link, when the link lies somewhere between the maximum gap position and the axis of misalignment. The critical region is shown in Fig. 4.7 by region Y. The stress has mean and alternating components, and, if a detailed stress analysis is required, their combined severity can be assessed by plotting them on a Goodman diagram (a graph of alternating stress versus mean stress), and comparing the resulting stress state with a limiting line for the membrane material.

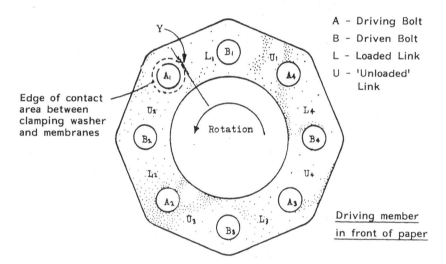

Fig 4.7   A typical ring-type membrane pack showing bolt connections and stressed membranes.

4.4   FORCES AND MOMENTS

4.4.1   Radial forces and vibrations

Lateral forces may arise due to misalignment or unbalance.

The forces due to misalignment result from the bending moments generated at a pack when it is angularly misaligned. Bending stiffness increases slightly with torque and axial displacement, but is chiefly determined by pack thickness. However, these forces and the corresponding moments due to angular misalignment are relatively insignificant, and are not likely to be of practical concern.

Radial forces due to unbalance are liable to be much more significant, particularly in any designs in which the membrane clamping bolts have to be disturbed during assembly and disassembly.

Balance may also drift in operation unless the clamping bolts are sufficiently large and tight to prevent pack movement altogether. If this latter problem is avoided, and if coupling designs are used in which the clamping bolts do not have to be disturbed for assembly, works balancing can be carried out to a good standard and should then be maintainable in service for long periods.

4.4.2 Axial forces and vibrations

The axial force generated by an axially displaced coupling is very largely due to the axial components of the link tensions, and is proportional to displacement, when torque is being transmitted, up to the point where the slack in the unloaded links is taken up. Up to this point the axial force also increases with torque. For still greater displacements the axial force increases faster and non-linearly, and the increase is independent of torque. The axial forces will not normally be difficult to accommodate, so long as the coupling is kept within its allowable axial displacement limits.

The two membranes of a spacer coupling are like springs, and the spacer will have a natural frequency of axial vibration corresponding to its mass and the spring stiffness.

The axial natural frequency covers a band due to the non-linear axial spring rate. The low end of the band is given by

$$\omega^2 = \frac{50 \cdot G_A}{M}$$

where

ω = natural frequency (Hz)

G_A = axial spring rate of one diaphragm assembly (kN/m)

M = spacer mass (kg)

It is advisable to keep this frequency above the maximum running speed of the coupled machines. In practice, any axial resonance is easily cured by a small change in the coupling stretch during installation.

4.5 FAILURE MODES

The principal failure modes of multiple membrane couplings are bolt failure, bending fatigue of the membranes where they emerge from between the clamping washers, or tearing of the membrane holes.

Bolt failures appear to have been the greater source of trouble in the past, and it is clear that bolt design and tightening are critical. Insufficient bolt tightness results in fluctuating shear and bending loads being applied to the bolts, which tends to cause fretting and early fatigue failure. Suitable design of the clamping washers can alleviate these problems, but the main solution appears to lie in carefully designing the bolts to carry high tensile loading, tightening them to high consistent extensions, and designing and sizing the whole coupling to keep the link tensions at a level which can be transmitted by friction alone at the bolts.

There will also remain some risk of bolt failure, however, and such failures could lead rapidly to catastrophic failure of the whole coupling due to flailing of the spacer, particularly with four bolt couplings. With designs using six or more bolts, the risk of catastrophic failures is reduced, since enough bolts could remain to transmit the drive for a short time, during which the high vibration level resulting from much increased coupling unbalance could be detected by a vibration pick up on the adjacent bearings, and could signal the need for unit shutdown. Undertightened bolts could also expose the membranes to tearing from the bolt hole to the adjacent outer periphery of the membrane, and this type of failure could also be sudden and catastrophic.

With adequate bolt design and tightness the most likely remaining cause of failure is fatigue of the membranes where they emerge from the clamping washers — point Y, Fig. 4.7. Since the outer membranes of the pack carry high pack bending stresses, however, failure occurs one membrane at a time, starting with the outer membranes, and consequently, there is again scope for using vibration to signal unit shutdown, since an outer membrane which broke at one end would probably buckle outwards about the other end under centrifugal force, and would then significantly increase the coupling unbalance.

It should be noted that fatigue failure of the membranes may be difficult to avoid totally with multiple membrane couplings, since there is still some uncertainty about the actual stresses in the membranes, and about the acceptable level of design stress, in view of the possible effect of inter-membrane fretting.

Couplings could also fail due to overtorque, for example, due to a bearing seizure, and in this case the failure would probably be by tearing of the membrane bolt holes at about four to six times normal torque.

To avoid serious damage and danger to personnel consequent upon any of these various possible failure modes, it is essential that an effective means is provided for retaining the coupling spacer, i.e., for preventing flailing in the event of a failure at either flexible element. The necessary degree of protection against flailing will depend strongly on the rapidity with which the unit can be brought to rest after the failure.

CHAPTER 5

Contoured Disc Couplings

5.1 DESCRIPTION

A typical contoured disc coupling is shown in Fig. 5.1. In a coupling of this type misalignments between the shafts of the coupled machines is carried by the deflection of thin flexible discs.

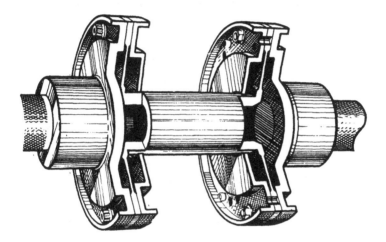

Fig 5.1   A contoured disc coupling.

The coupling has a central spacer tube with flexible discs at each end. The tube is connected to the inside diameter of the discs while the outside diameter is attached to a flange connected to the hub and shaft of the coupled machines. The disc is thin and has to be protected from damage and the coupling is fitted with shroud plates to contain the flexible discs inside a partially enclosed chamber at each end.

The discs are machined from high quality forgings of various high strength materials, and transmit torque in shear between the inner and outer rims, whilst accommodating misalignments by flexure. The discs can be combined in various ways to form couplings, but in the simplest arrangement one is used at each end of a spacer tube.

The cross section through a typical disc is shown in Fig. 5.2 and it consists of inner and outer flanges connected together by a flexible diaphragm of variable cross section.

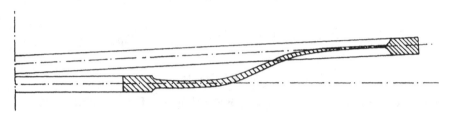

**Fig 5.2   The deflection of a contoured disc under combined
angular and axial misalignment.**

The fundamental limitation on the performance of disc couplings is
fatigue of the disc under the combination of stresses due to torque, centrifugal effects, and axial and angular misalignment. The type of profile
affects the distribution of the stresses and the location of the critical stress
condition. The most favourable distribution appears to be achieved with a
hyperbolic profile, in which the thickness is inversely proportional to the
radius squared. This gives uniform shear stress due to torque, and the
critical overall stress condition then occurs at the outer radius of the disc.
The ratio of the inner to outer disc radii also affects the stress pattern, and a
ratio of 0.52 is commonly used, giving a reasonable compromise between
several design criteria.

5.2   APPLICATION

The couplings are of a very simple shape and can, therefore, be balanced to
high standards, and are particularly suitable for high speed applications.
The contoured discs, depending on their design, can allow angular deflections of up to 0.5 degrees, and are of low stiffness, so that they only apply
low loads to the coupled machines. This type of coupling is, however, of a
relatively large diameter, particularly when increased axial deflections are
required.

Figure 5.3 gives guidance on the power and speed capacity of contoured
disc flexible couplings of various outside diameters. For specified design
stress levels, increasing the diameter increases the torque and misalignment capacity, whilst increasing the disc thickness increases torque capacity but reduces misalignment capacity. The diameter is ultimately limited
by centrifugal stress. Torque capacity is limited by torsional buckling of the
disc, but the discs are designed to keep the shear stresses in continuous
operation below 30 per cent of that necessary to cause buckling.

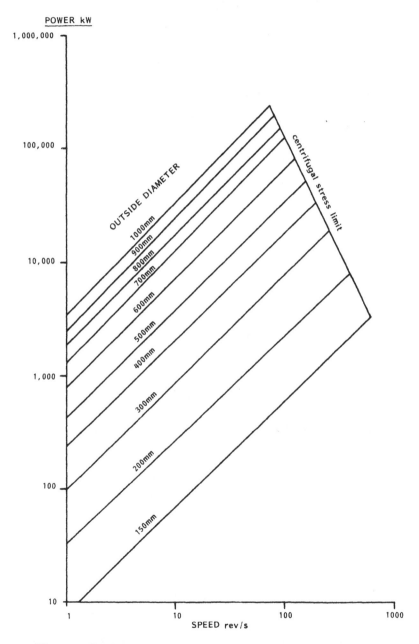

Fig 5.3 Performance envelope for contoured disc couplings.

5.3   PERMISSIBLE MISALIGNMENT

Each contoured disc is capable of allowing angular misalignments of up to about 0.009 radians, and axial displacements either side of the mean position of up to about 0.01 of the disc outside diameter. These values correspond broadly to the performance of the thinnest of the range of discs offered by vendors and, therefore, to the lowest range of torque capacity. Vendors offer a choice of up to three disc thicknesses and for the thickest discs, giving the maximum torque capacity, the allowable angular misalignments and axial deflections are about half the above values.

5.4   FORCES AND MOMENTS

Because the discs are of low stiffness in terms of angular misalignment the radial forces and moments applied to the coupled machines are very low and unlikely to be of practical concern.

The axial forces generated by axial displacement are also low, but increase when the couplings are deflected towards their maximum value. However, the forces involved are well within the capacity of machine thrust bearings.

Since the couplings are effectively an elastic component the various angular and axial stiffnesses are easy to measure and calculate, and the vendors can, therefore, provide accurate information on the forces that will be generated by any particular coupling.

5.5   BALANCING AND CENTERING

Lateral forces due to unbalance can usually be kept low with this type of coupling because of the inherent symmetry and total absence of backlash in the design, and because the bolting at the outer flange does not have to carry high stress so that relative flange movements can be readily prevented, and good balance can be permanently retained. The effect of the coupling on lateral vibrations of the shaft overhangs can be more significant, however, because the main mass of disc couplings is between the hubs and, therefore, more overhung relative to the machine bearings than in the case of some other couplings.

5.6   COOLING

Because this type of coupling is of relatively large diameter it can generate high windage losses and sometimes noise. Any coupling guards should not, therefore, be fitted too close to the coupling. Vendors should be asked for advice on whether air cooling arrangements might be necessary and on the optimum design of guards and housings.

5.7 FAILURE MODES

One of the merits of disc couplings is that the various stresses are determinable with reasonable accuracy, and are kept within conservative limits during all rated operating conditions. Consequently failures are unlikely except due to various abnormal causes such as machine seizure, extended operation with misalignments in excess of design limits, or damage to the discs due to mishandling.

The mode of failure due to excessive overload would be torsional buckling, and then rupture, of the disc, and this limit is designed to be at about 330 percent of normal rated torque. The problem can be overcome by including a gear coupling at each mesh designed with just enough backlash to take over the load at the buckling limit.

The mode of failure as a result of excessive misalignment or handling damage will be by fatigue, and the ultimate failure due to fatigue could well occur quite suddenly. A fail-safe arrangement is therefore considered desirable, and the guard plates used to enclose the spacer tube and discs should meet this requirement.

CHAPTER 6

Elastomeric Element Couplings

6.1 DESCRIPTION

Elastomeric element couplings use natural or synthetic rubber elements, loaded in shear or compression. They are manufactured in a wide range of sizes and in a seemingly endless variety of styles. The selection of a coupling for a particular application depends on the required torque, torsional flexibility, damping, and allowances for shaft misalignment. Nearly all couplings consist of three basic assemblies; two shaft hubs and a flexible element connecting them.

The more widely used types of smaller elastomeric couplings are the spider, the doughnut, the pin & bush, and the tyre coupling (Fig. 6.1). For low power and non-critical applications, each user may have his own preference based on cost, flexibility, serviceability, and acceptable reliability. The ISO range of pumps is a typical example of such applications.

Doughnut

Spider

Tyre

Pin & bush

Fig 6.1 Types of elastomeric coupling

In general, for larger machines subject to shock loads, a rubber-in-compression coupling is preferred for long term, maintenance-free operation; the absence of metal to metal contact eliminates the need for lubrication. These couplings are also preferred when machines must run, for a limited time at least, even after failure of the rubber elements. Rubber-in-shear couplings, whilst providing low torsional stiffness, do not continue to drive after such failure, and should be avoided if this is an important criterion.

One type of rubber-in-compression coupling is illustrated in Fig. 6.2. The rubber elements are usually wedge shaped, but when a torsionally softer coupling is required, or where axial loads need to be reduced, round elements can be substituted, albeit with a generally reduced maximum torque rating.

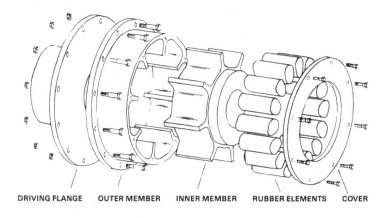

DRIVING FLANGE   OUTER MEMBER   INNER MEMBER   RUBBER ELEMENTS   COVER

Fig 6.2   Rubber-in-compression coupling.

Due to the blade design and the element design substantial axial, radial, and angular misalignments can be accommodated without metallic contact.

The wedge elements are assembled with 6–8 per cent pre-compression and in this way are never unloaded, even at peak torque rating in either direction. Backlash effects under impulse loads are thus prevented, eliminating an important source of noise, wear, and impact damage found in some metallic couplings.

Because of the element pre-compression it is essential that a lubricant is applied to coupling cavity walls and to the elements before assembly. Without this essential lubrication, fitting is difficult and may result in damage to the blocks. Silicone fluid between 300 and 1000 Cp is a suitable lubricant for all rubbers, as is silicone grease, soft soap, and some detergents, but neoprene and nitrile rubbers may alternatively be lubricated with mineral lubricating oils.

Coupling parts are typically manufactured from steel castings or steel forgings where higher than normal peripheral speeds are encountered.

A limited-end-float feature is available for driving or driven machines not fitted with an axial location bearing. Generally, self-lubricated thrust pads are used, but if the axial load is significant, then an alternative annular thrust ring system may be incorporated. Cardan shaft couplings with rubber end stops, and spacer couplings are also available (Fig. 6.3).

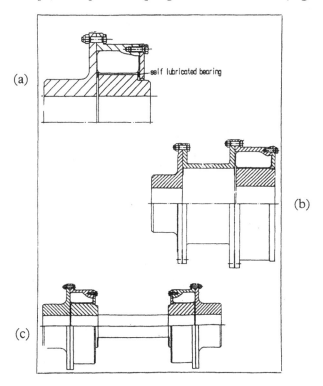

Fig 6.3 (a) Limited end float couplings (b) Spacer coupling (c) Cardan shaft coupling.

The coupling, of course, may be split to allow free rotation of each machine for maintenance, by removing only the flange bolts. Couplings which require a mechanical method of separating the outer member from the driving flange are fitted with tapped jacking holes in the cover. During maintenance the cover can be removed in situ allowing examination and, if necessary, replacement of the rubber elements. For special applications and for couplings in confined spaces, the cover can be split, which allows for easy removal over a shaft. Final shaft alignment measurements (see Chapter 8) are normally measured with the coupling fully assembled.

Although many resilient couplings are selected on no more information than shaft size, power, and speed, potential problems can be avoided at an early stage if ISO standard 4863–1984, *Resilient shaft couplings – information to be supplied by users and manufacturers*, is used as a check list, and as a basis for specifying the required characteristics.

For couplings in accordance with standard API 671 the use of API 671 data sheet is most convenient. The data sheets in Appendix II are also recommended for other couplings.

6.2 APPLICATION

Elastomeric couplings are the obvious choice where torsional critical speeds, or torsional cyclic stresses need to be much reduced. They are, therefore, commonly used for reciprocating machines. A typical speed-power application envelope is shown in Fig. 6.4 for rubber-in-compression industrial elastomeric couplings. The application limits are determined by the elastomer stress due to torque, and by centrifugal forces acting on the metal parts. Torque capacity is proportional to the coupling diameter and, hence, to centrifugal forces, which give a performance limit for power and speed. Speed is most often limited by the allowable peripheral speed of the outer diameter of the coupling body, and a typical value of 60m/s is quoted for cast steel. Elastomeric couplings operating above 60m/s are constructed of high integrity cast steel or forged steel, and are often stiffened by the use of bearings, allowing torsional deflection only.

For this range of couplings standard sizes are available up to 1 kW/r/s and special designs are manufactured up to 17 kW/r/s.

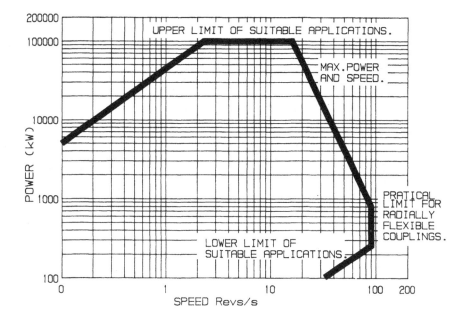

Fig 6.4   Power-speed application boundary.

6.3   PERMISSIBLE MISALIGNMENT

Figures are quoted in manufacturers' literature for the axial, radial, and angular misalignments with which the coupling can cope. However, a close study should be made to see if these are quoted as separate maximum values (i.e., assuming that the other two possible misalignments are in fact perfect) or maximum values quoted with the three types of misalignment present simultaneously. For long coupling life, bearing life, and seal life, shafts must be aligned as accurately as practicable. Coupling rubber elements will suffer serious deterioration if accurate alignment is not carried out, as wear on the elements is proportional to speed, operating time, and degree of misalignment. Indications of allowable values of misalignment are given in Figs 6.5 and 6.6.

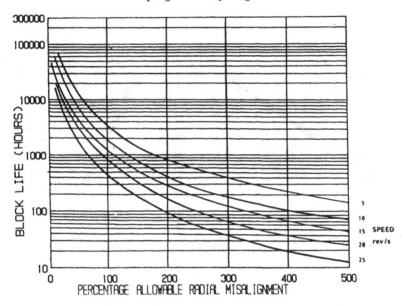

Fig 6.5   Block life vs misalignment.

No graph can ideally reproduce all the factors caused by misalignment, but Fig. 6.5 gives an indication of element life against allowable radial

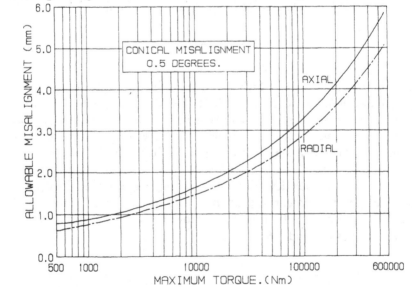

Fig 6.6   Misalignment vs maximum torque.

misalignment for a rubber in compression coupling. Whilst typical published data, Fig. 6.6 gives maximum possible values with all three types of misalignment occurring simultaneously, a general value of 25 per cent of the maximum values should not be exceeded for acceptable block life.

Axial, radial, or angular misalignment will induce bearing loads of a magnitude dependant on the axial, radial, or angular stiffness, as explained in section 6.4 below. In some circumstances these forces may be so large, or in such an unfavourable direction, that the alignment must be improved. In these cases it is the bearings and not the coupling, which dictate the maximum acceptable misalignment.

6.4 FORCES AND MOMENTS

Both the radial and axial stiffness values of a coupling can be found by laboratory testing as well as by theoretical analysis. However, it is often useful to be able to predict these values using the following formula in the case of special designs.

Radial stiffness, K_r

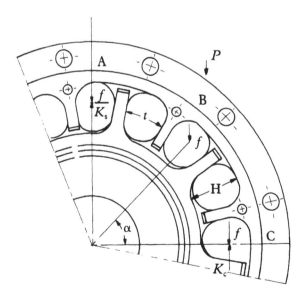

K_r at A is considered as pure shear.
K_r at C is considered as pure compression.

Consider coupling element spaced between these two extremes at B.

$$f = f_c \cos \alpha + f_s \sin \alpha$$
$$x_c = x \cos \alpha; \quad x_s = x \sin \alpha$$
$$f_c = K_c x_c; \quad f_s = K_s x_s$$
$$P = K_c \times \cos^2 \alpha + \sin^2 \alpha \times K_s$$
$$K_r = \frac{P}{x} = \Sigma (K_c \cos^2 \alpha + K_s \sin^2 \alpha)$$

$$K_s = \frac{GA}{t}; \quad K_c = \frac{E_c A}{t}$$

The shear and compressive module can be combined with a shape factor to give a constant E_R applicable to a particular form of coupling giving

$$K_r = \frac{W_c H_c n_c E_R}{t_c}$$

Where
   W_c = Cavity length (mm)
   H_c = Cavity height (mm)
   n_c = number of cavities
   t_c = Chordal pitch (mm)
   f = Force on one element (N)
   x = Deflection (mm)
   P = Total force (N)
   K_s = Shear stiffness (N/mm)
   K_c = Compressive stiffness (N/mm)
   A = Area (mm^2)
   G = Shear modulus (N/mm^2)
   E_c = Compressive modulus (N/mm^2)

As an example, E_R varies between 0.28 kg/m^2 for rubber hardness 50 to 1.12 kg/m^2 for rubber hardness 80, for a typical wedge type block, but would be significantly less for a round block.

Axial stiffness, K_A

$$K_A = \frac{W_c H_c n_c E_A}{t_c} \text{ (N/mm)}$$

E_A = A constant dependent on the shear modulus and shape factor.

As an example, E_A varies between 0.05 kg/m^2, for rubber hardness 50 to 0.37 kg/m^2 for rubber hardness 80 for a typical wedge type block, but would be significantly less for a round block.

Moments

Consider one coupling

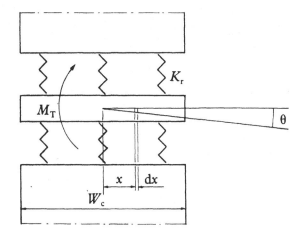

Where

$\dfrac{M_T}{\theta}$ = bending or conical stiffness

K_r = coupling radial stiffness

W_c = cavity length

Neglecting effects of axial stiffness

$$M_T = \theta \int_{-W_c/2}^{W_c/2} K_r x^2 \, \frac{dx}{W_c}$$

$$\frac{M_T}{\theta} = \frac{K_r W_c^2}{12}$$

With axial stiffness included.

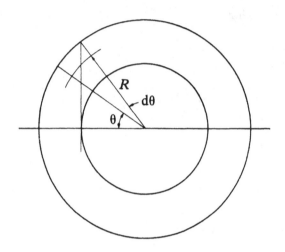

R = mean element radius

$$\frac{M_T}{\theta} = 4 \int_0^{\pi/2} R^2 \cos^2\theta \frac{K_A}{2\pi} \, d\theta$$

$$= \frac{R^2 K_A}{\pi} [\sin\delta\theta \, \cos\theta + \theta]_0^{\pi/2}$$

$$\frac{M_T}{\theta} = \frac{K_A R^2}{2}$$

Therefore, the transverse bending stiffness of one coupling

$$\frac{M_T}{\theta} = \frac{K_r W_c^2}{12} + \frac{K_A R^2}{2}$$

For a cardan shaft, then, the following applies:

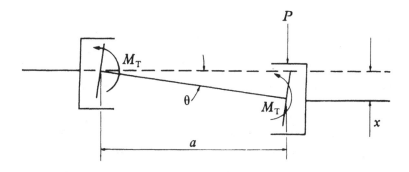

For two couplings

$$\frac{M_r}{\theta} = \frac{P_a}{\theta} = 2\frac{K_r W_c^2}{12} + \frac{K_A R^2}{2}$$

$$= \frac{K_r W_c^2}{6} + K_A R^2 \tag{1}$$

For shaft end radial stiffness

$x = a\theta$ (for small angles)
$\theta = x/a$

Substituting into (1)

$$\frac{P_a^2}{x} = \frac{K_r W_c^2}{6} + K_A R^2$$

$$P = x\left[\frac{K_r W_c^2}{6a^2} + \frac{K_A R^2}{a^2}\right]$$

6.5 BALANCING

For rubber-in-compression couplings operating at relatively low speeds, the machining tolerances are usually sufficient for parts not to warrant balancing. However, for couplings operating at above 80 per cent of the published maximum speed, balancing should be carried out. A coupling balance quality grade in accordance with ISO standard 1940 is suitable for elastomeric couplings, and for most assemblies G16 is normally specified (Table 6.1). For couplings in the process plant industry G6.3 is preferred, and under extreme circumstances G2.5 may be stipulated, although this will probably require closer coupling tolerances. Where G6.3 or G2.5 is required, weight balancing of the elements may be required, with each set clearly identified. In this way matched pairs can be fitted into opposite cavities at initial assembly and, if required, during subsequent overhaul. Weight balancing of the fitted bolts may also be carried out.

Table 6.1 Balance quality grade. ISO standard 1940

Balance quality grade G	$e\omega$* (mm/s)	Rotor types — General examples
G 4 000	4 000	Crankshaft-drives of rigidly mounted slow marine diesel engines with uneven number of cylinders.
G 1 600	1 600	Crankshaft-drives of rigidly mounted large two-cycle engines.
G 630	630	Crankshaft-drives of rigidly mounted large four-cycle engines. Crankshaft-drives of elastically mounted marine diesel engines.
G 250	250	Crankshaft-drives of rigidly mounted fast four-cylinder diesel engines.
G 100	100	Crankshaft-drives of fast diesel engines with six or more cylinders. Complete engines (gasoline or diesel) for cars, trucks and locomotives.
G 40	40	Car wheels, wheel rims, wheel sets, drive shafts. Crankshaft-drives of elastically mounted fast four-cycle engines (gasoline or diesel) with six or more cylinders. Crankshaft-drives for engines of cars, trucks and locomotives.

(Continued)

Table 6.1 continued   Balance quality grade. ISO standard 1940

Balance quality grade G	$e\omega$ * (mm/s)	Rotor types General examples
G 16	16	Drive shafts (propeller shafts, cardan shafts) with special requirements. Parts of crushing machinery. Parts of agricultural machinery. Individual components of engines (gasoline or diesel) for cars, trucks and locomotives. Crankshaft-drives of engines with six or more cylinders under special requirements.
G 6.3	6.3	Parts or process plant machines. Marine main turbine gears (merchant service). Centrifuge drums. Fans. Assembled aircraft gas turbine rotors. Fly wheels. Pump impellers. Machine-tool and general machinery parts. Normal electrical armatures. Individual components of engines under special requirements.
G 2.5	2.5	Gas and steam turbines, including marine main turbines (merchant service). Rigid turbo-generator rotors. Rotors. Turbo-compressors. Machine-tool drives. Medium and large electrical armature with special requirements. Small electrical armatures. Turbine-driven pumps.
G 1	1	Tape recorder and phonograph (gramophone) drives. Grinding-machine drives. Small electrical armatures with special requirements.
G 0.4	0.4	Spindles, disks, and armatures of precision grinders. Gyroscopes.

*$\omega = 2\pi \times N/60 \propto n/10$, if n is measured in revolutions per minute and ω in radians per second. e is the eccentricity of the centre of gravity.

Most coupling designs are short, i.e., axial length is usually less than half the diameter, and the mass distribution is such that single plane balancing of each half-coupling assembly is adequate for the majority of applications. Two-plane balancing is necessary for long coupling arrangements, such as cardan shafts or shaft systems where the support bearings are axially close (less than twice coupling length).

Balancing of each component is carried out on special mandrels; inner member, outer member, and cover (if necessary) for a half coupling, and the driving flange for a full coupling. Outer member and cover are then balanced together and correlated, then fitted to the driving flange with fitted bolts, balanced again, and correlated.

Cardan shaft couplings and spacer couplings are similarly component balanced, correlated, and assembly check-balanced. For a more detailed discussion of balancing requirements the reader should refer to Chapter 9, section 9.3.

6.6   TYPES OF RUBBER

A wide variety of rubbers are used in elastomeric couplings, e.g., natural, styrene–butadiene, neoprene, and nitrile. Polyurethene is limited at present to the smaller couplings, where it can increase the torque rating by up to a factor of three, but severely reduces the torsional flexibility.

To select the most suitable rubber the following information is required.

(a)   What is the highest temperature likely to be encountered in service by the coupling?

(b)   What is the highest temperature at which continuous service will be required?

(c)   What is the lowest temperature at which the coupling must remain operable?

(d)   What fluids will be encountered in service (intermittent, occasional, etc.) and at what temperature?

(e)   Is long term weather or ozone resistance an important factor?

Particular attention should be paid to the running temperature of the rubber blocks. This may be torsionally augmented, and can also be significantly affected if the coupling is protected with anything other than a mesh type guard, especially if it is closely confined in the interests of silencing the machine. In an extreme case, some form of cooling may be required.

For a list of rubber materials, their general properties and relative merits, reference should be made to Table 6.2 and the following notes. The dynamic behaviour of rubber (stiffness and damping properties) are discussed in section 6.8 below.

Table 6.2

Rubber compound General characteristics	Natural	Styrene– butadiene	Neoprene	Nitrile
Resistance to compression set	Good	Good	Fair	Good
Resistance to flexing	Excellent	Good	Good	Good
Resistance to cutting	Excellent	Fair	Good	Good
Resistance to abrasion	Excellent	Good	Good	Good
Resistance to oxidation	Fair	Fair	Very good	Good
Resistance to oil and petroleum	Poor	Poor	Excellent	Excellent
Resistance to acids	Good	Good	Fair	Fair·
Resistance to water swelling	Good	Good	Good	Good
Service temperature (max. continuous)	80°C	100°C	100°C	120°C
Service temperature (minimum)	—50°C	—40°C	—30°C	—40°C

Natural rubber (NR)
Natural rubber has good mechanical properties, e.g., tensile strength ($24\,000$ kN/m^2) and extensibility, elasticity, and resilience (and, hence, low hysteresis and heat build up). It is a good all-round material, but is not very resistant to oxidation and ozone cracking and, therefore, does not age well at high temperatures, although it performs well at low temperatures.

Styrene–butadiene (SBR)
Styrene–butadiene rubber is the most widely used rubber in the world today. For large rubber-in-compression couplings it is the generally preferred material. When compared with natural rubber it has inferior gum tensile strength, but when reinforced with carbon black, strengths of greater than $21\,000$ kN/m^2 can be achieved.

Chloroprene (CR)
For example, neoprene. As well as having high gum strength, these rubbers also have good oil and chemical resistance. Because of the chlorine in the molecule, CR has inherent flame resistance and its products can be made self-extinguishing. This rubber is the first choice for the coal mining industry.

Nitrile

Generally, these rubbers have good oil, petol, and solvent resistance and are also resistant to dilute acids and alkalis. They also exhibit moderate to good temperature resistance.

Rubber hardness grades

Once a rubber has been selected, a hardness value must be determined. Hardness is measured in degrees on the IRHD scale or the Shore A scale. Values are similar, but the Shore A readings are usually 1–3 degrees higher than the IRHD readings. Hardness values are normally based on a nominal figure ± 5 degrees, e.g., 50 ± 5 degrees, or as a hardness range 50–60 degrees. For torsionally insensitive couplings, a hardness value of 70 is preferred, but when torsional requirements (see section 6.8) dictate the rubber grade, then values between 50 and 80 may be chosen.

Storage of spare rubber blocks

The British Standards Institution recommends the following storage conditions for vulcanised elastomers, in order to avoid premature ageing and consequent changes in properties. Ageing is indicated by softening, hardening, cracking or other surface degradation. Deterioration can be accelerated by the synergistic effect of storage conditions such as heat, light, humidity, oxygen, and ozone. Ageing can be minimized by careful attention to storage conditions.

Temperature. The storage temperature should be less than 25°C. For long term storage (greater than three years) a 15°C maximum is recommended.

Humidity. Moist conditions should be avoided, and no condensation should occur.

Light. All rubber elements should be protected against light, especially sunlight and artifical light with ultra-violet content.

Oxygen and ozone. Components should be protected from air circulation by storage in containers. Ozone is particularly damaging and, for this reason, parts should not be stored near electrical equipment where sparks are possible.

Deformation. Rubber elements should be stored in a strain-free condition.

Liquids. Rubber elements should not be contaminated with liquids.

Cleaning. If cleaning is required, soap and water is relatively harmless. The washed components should be allowed to dry at room temperature.

Storage time. Where items have been stored for five or more years, a laboratory check on quality must be carried out before use.

6.7 FAILURE MODES

A rubber-in-compression coupling operating in a correctly-assessed system requires a minimum of maintenance, but periodic inspection is recommended. In particularly arduous duties (e.g., reciprocating machines) it is recommended that the elements are withdrawn and inspected every 5000–10 000 operating hours to assess their remaining life. Provided the stress and strain limitations of the rubber are observed, and the environmental conditions are favourable, then the blocks are capable of a very long life; 10 to 15 years is not unusual.

Inspection should determine whether any damage or change has occurred and, in cases of advancing deterioration, arrangements should be made to renew the elements as soon as convenient.

The following are failure modes to look for.

Abrasion
Rubber dust in small amounts is normal, but large quantities indicate a need for lubrication. In this case both elements and cavity surfaces should be coated with silicone fluid. Elements so abraded that they are loose in the cavities should be changed as soon as possible, as the rate of abrasion of loose blocks increases rapidly.

Cutting
Deeply cut elements indicate the application of excessive torque loadings, and these blocks should be changed.

Deformation
Elements greatly distorted when unloaded indicate operation under hot conditions, and should be replaced.

Rubber deterioration
Deterioration of rubber is shown by crumbly texture, brittleness, surface cracking, or stickiness. Any of these indicate that the rubber is undergoing change, and therefore the elements should be replaced.

Rubber hardness change
On torsionally critical applications, rubber blocks which have hardened or softened more than 10 degrees (IRHD or Shore A), from the hardness

number shown on the identification mark, should be replaced as the rotor critical speed, and its damping, will be significantly changed from their original values.

6.8 TORSIONAL CHARACTERISTICS

Both the torsional stiffness and the torsional damping characteristics of a given coupling depend on the rubber grade (IRHD hardness), the rubber material, and the geometry of the rubber elements. For general industrial applications, 60 or 70 IRHD hardness wedge shaped elements of styrene–butadiene are normally preferred. If a transient analysis or torsional analysis is carried out (see Chapter 9) then the determination of the torsional response may indicate a different hardness or a different element type and, in some cases, a change to a different material. However, if the response characteristics of the system cannot be determined with certainty at an early stage, then a modification, often on site, can easily be effected by changing the elements.

Decreasing hardness gives decreasing stiffness, and in certain applications a hardness value of 45 is permissible. For lower stiffness values and reduced axial thrust, the wedge elements can be changed, generally to a round element. This may, however, reduce the maximum torque rating.

Dynamic torsional stiffness

The dynamic torsional stiffness is the ratio of the vibratory torque to the deflected angle during one vibration cycle about the mean position of torque and the mean position of deflection. Dynamic values vary for different rubber grades and materials and for different values of normal torque due to the non linear nature of the static torque deflection curve (illustrated for styrene–butadiene in Fig. 6.7) and are often published at values of 25, 50, 75, and 100 per cent of nominal torque. Figure 6.8 illustrates these values, again for styrene–butadiene rubber obtained with a vibratory torque amplitude of 12.5 per cent of maximum torque, a temperature of 30°C, and at a reference frequency of 10 Hz.

Whilst dynamic torsional stiffness depends mainly on mean torque and mean angle; frequency, vibratory torque amplitude, and temperature must also be taken in account if significantly different from the reference values. Design values are obtained from vibration tests on various materials and grades to obtain these correction factors. Figures 6.9 and 6.10 show how the stiffness factor varies with temperature and frequency for styrene–butadiene rubber.

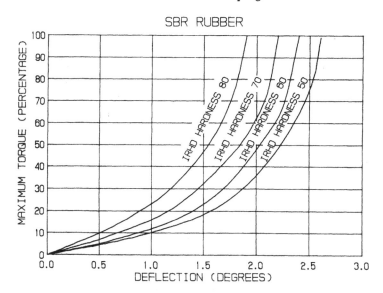

Fig 6.7   Coupling torque deflection (SBR rubber).

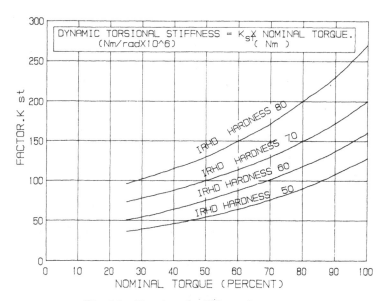

Fig 6.8   Torsional stiffness factor.

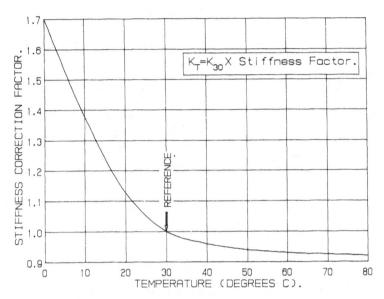

Fig 6.9   Stiffness factor vs temperature.

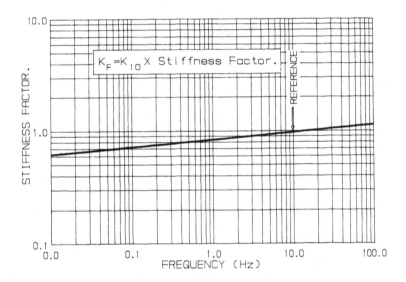

Fig 6.10   Stiffness correction factor vs frequency.

Torsional damping

In a shaft system which is being torsionally excited, damping limits the vibratory displacements and stresses, particularly at rotational speeds equal to the torsional natural frequencies, whether occurring within the operating speed range or during run-up and run-down.

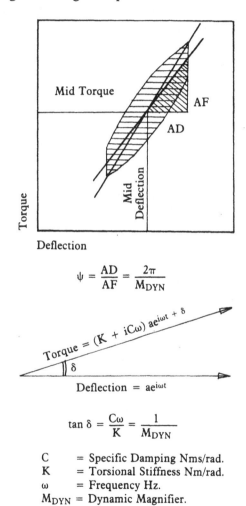

$$\psi = \frac{AD}{AF} = \frac{2\pi}{M_{DYN}}$$

$$\tan \delta = \frac{C\omega}{K} = \frac{1}{M_{DYN}}$$

C   = Specific Damping Nms/rad.
K   = Torsional Stiffness Nm/rad.
ω   = Frequency Hz.
M_{DYN} = Dynamic Magnifier.
δ   = Phase Angle rad.
ψ   = Damping Energy Ratio.

Fig 6.11

The nominal static torque deflection curve at the known operating condition, shown as mid-deflection and mid-torque (Fig. 6.11), under a condition of vibratory torque, changes to the hysteresis curve shown as an ellipse in the figure. The area of the ellipse, (AD) and the area of the triangle (AF) can then be expressed as a ratio which is defined as the damping energy ratio AD/AF or ψ. This ratio is also expressed as $2\pi/M_{DYN}$ where M_{DYN} is the dynamic magnifier.

Coupling damping varies directly with torsional stiffness and inversely with frequency, for a given rubber material and grade and this relationship Cw/K is equal to the inverse of the magnifier, or tan δ.

The values of dynamic magnifier for different rubbers and grades is illustrated in Fig. 6.12 for a reference frequency of 10Hz. A further graph is then required to correct the dynamic magnifier for frequency and operating temperature and this is shown in Fig. 6.13. The corrected value of magnifier can then be used in torsional calculations, as explained in Chapter 9.

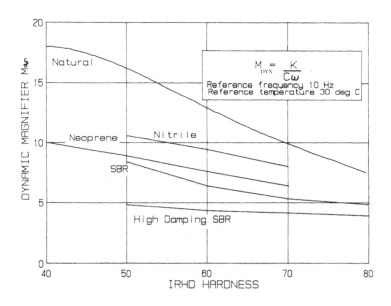

Fig 6.12   Damping characteristics

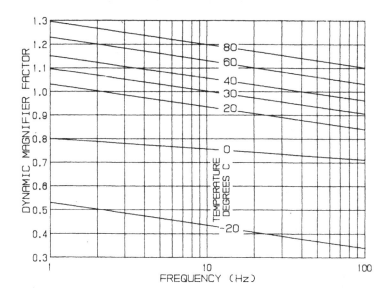

Fig 6.13   Dynamic magnifier vs frequency

CHAPTER 7

Quill Shaft Couplings

7.1 DESCRIPTION

A typical quill shaft coupling is shown in Fig. 7.1. The central quill shaft is flexible in bending and can, therefore, absorb some angular and lateral misalignment between the coupled machines. The quill shaft is also torsionally flexible, but it is axially rigid.

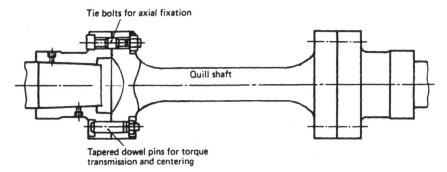

Tie bolts for axial fixation

Quill shaft

Tapered dowel pins for torque
transmission and centering

Fig 7.1   A typical quill shaft coupling

The end connections of quill shafts are a critical feature of the design because they represent the transition from the relatively flexible quill member to the relatively rigid connections to the drive shafts of the coupled machines. The fillet radii at the ends need to have controlled profiles and surface finish, and the shaft material quality must be very high. Flanged connections are frequently used to enable the quill shaft to be fitted or removed easily, without disturbing the coupled machines. Ideally the hubs should be hydraulically fitted to the machine drive shafts, to obtain a drive system of minimum weight and of generally uniform cross section.

7.2 APPLICATION

Quill shaft couplings are advocated by a few compressor and gear manufacturers. The simplicity of the design, its freedom from lubrication and maintenance, inherently good balance, and susceptibility to precise analysis are very desirable features of a quill shaft coupling. Couplings with increased capability for allowing angular and lateral misalignment will, however, also have increased torsional compliance. Quill shaft couplings need, therefore, to be designed as an integral part of the complete machine train if the optimum dynamic behaviour is to be obtained from the system. This requires detailed engineering liaison between the manufacturers of the coupled machines. If, therefore, the buyer of a machine train wishes quill shaft couplings to be considered, it is, in practice, vital to state this as a firm requirement at the earliest opportunity.

The axial rigidity of the couplings also opens up opportunities for improved layouts of machine systems. Machines which generate axial loads can sometimes be arranged so that the loads are partially balanced across the quill shaft connection. In some systems, a quill shaft coupling can eliminate the thrust bearing in the gearbox or electric motor. In geared drives, with gears designed to transmit thrust, the thrust bearing, which locates the machine rotors axially and carries all residual thrust, is preferably located on the low-speed shaft in order to reduce bearing losses.

The effect of the torsional compliance of the coupling, on the resonant torsional frequency of the machine system, needs to be considered. It can provide a useful means of lowering this frequency and of changing nodal positions. The torsional compliance can also be used to reduce the effect of impulsive torques in a machine system.

Quill shaft couplings are particularly appropriate for high speed machines because of their excellent inherent balance, and there are no intrinsic limits on their capability in terms of power and speed. Practical limits arise from the maximum misalignment which can be permitted for a particular coupling, within the allowable fatigue stresses for its material. The couplings are also of a relatively low mass compared with other types and, therefore, produce the minimum reduction in the lateral critical speed of coupled machines, for which they constitute an overhung mass at the end of the shaft.

The optimum application of quill shaft couplings really requires their individual design to match a particular machine drive train, since this provides an opportunity to obtain the best matching between the required lateral and torsional flexibility, consistent with an infinite fatigue life.

7.3 OPERATING LIMITS

The permissable angular and lateral misalignment depends on the detailed design of the quill shaft and its ability to carry the resultant fluctuating bending stresses, superimposed on the torsional stress from the drive torque, without fatigue failure occurring. The stress levels existing are amenable to calculation and measurement, and need to be kept well below the fatigue limit because of the very large number of bending cycles to which a quill shaft will be subjected in service.

The bending moments generated in the quill shaft by angular misalignment will give rise to moments and forces, which will be applied to the coupled machines. These can be calculated from the quill shaft dimensions and the misalignment patterns of the machines.

A disadvantage of the quill shaft coupling is that the distance between shaft ends may need to be increased, in order to allow a practical degree of misalignment. Some gear manufacturers, however, offer hollow gear and pinion shafts with long quill shafts fitted inside (driven from the normally free ends of the shafts), thus achieving a compact shaft layout, adequate misalignment capability, and, possibly, the elimination of a thrust bearing as well.

Quill shafts will be designed to operate without fatigue failure within an envelope of allowable lateral and angular misalignment and transmitted torque. If failure does occur it would be most likely to arise in the transition region between the simple cylindrical portion of the shaft and the end coupling flanges. Any fatigue cracks should start at the outer surface since this is the most highly stressed region and should, therefore, be detectable by most conventional crack detection methods.

CHAPTER 8

Alignment

8.1 ALIGNMENT CHANGES DURING OPERATION

For long-term trouble-free performance of machines, adequate alignment at normal operating conditions is essential. Without this, there will be more vibration, and the natural life of coupling components, machine bearings, and seals will be curtailed. Starting, stopping, process changes, and upsets all cause inevitable displacements of the shafts; in industry, flexible couplings are almost universally used to accommodate such movements, thus avoiding the formidable difficulties which would occur with solid couplings. Flexible couplings reduce the stressess at times of alignment change but, for high-speed process machines, they are not an alternative to adequate operating alignment.

In general, the alignment measured during shut-down, when the machines are cold, will gradually change as they reach operating conditions. For precise operating alignment, one should, therefore, predict the changes, and align the shafts with corresponding offsets, such that these reduce to zero during operation. The rest of this chapter is a guide to predicting alignment changes, and hence the offsets which these changes require. It should be noted, however, that the magnitude of the changes is such that most motor-driven machines operate successfully when aligned without offsets. In contrast, steam and gas turbines, pumps and compressors operating at unusually high or low temperatures, and machines with journal bearings larger than 100 mm diameter, require particular consideration. For gear couplings, note also that a perfect operating alignment is to be avoided; such an alignment does not allow an adequate lubricant film to be established between the mating teeth. A minimum misalignment of 0.00075 radians should be maintained at each end of the spacer at all times (see section 3.2.2).

The particular shut-down conditions, at which the alignment is measured, should be considered before quantifying alignment changes. Measurement is usually at ambient temperature, but some machines have alignment checks at standby temperature (e.g., with heated lubricating oil circulating in preparation for start-up), while others may have process fluid circulating through them at operating temperature. There is also 'hot alignment', done as soon after shut-down as can be managed. Which of these conditions is chosen is largely a matter of convenience, but the

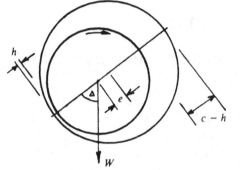

c = clearance
h = oil film thickness
e = eccentricity
Δ = attitude angle
ε = eccentricity ratio

$$= \frac{2e}{c} = 1 - \frac{2h}{c}$$

Note:

$\varepsilon = 0$ when shaft is concentric
with bearing
$\varepsilon = 1$ when shaft is in contact
with bearing

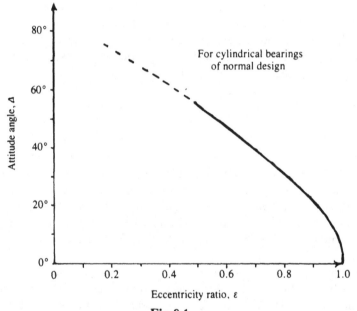

Fig 8.1

condition should be identified, and the ambient temperature or other relevant condition should be recorded.

Shaft displacements occur, particularly during start-up and shut-down, for various reasons:

- thermal expansion of bearing supports, and of shaft lengths;
- change in radial or axial forces – magnitude or direction;
- change in thickness of the bearing oil-film;
- change in piping forces;
- change in baseplate support.

The significance of each of these is considered in the following paragraphs.

A steam or gas turbine driving a cold pump is an obvious example of a combination which may require an offset alignment; during start-up the turbine frame will expand upwards more than the pump body, and the shafts will move radially relative to each other. For calculating expansions use a coefficient of 11×10^{-6} per unit length per degree Celsius for ferritic materials, and 16×10^{-6} for austenitic materials. Assumed ambient temperatures should be recorded so that the predicted growth is easily revised if alignment is measured at a significantly different temperature.

Thermal growth in the horizontal plane is often overlooked, although for large gearboxes it may require that both driver and driven machine are horizontally offset from the gear. The growth in shaft–centre distance is sometimes allotted equally to each shaft but, knowing the location of the dowels, or the fitted holding-down bolts, the total horizontal growth can be allotted more rationally.

Estimating the effective operating temperature of a bearing support is not easy, as the bearing will be hot, and the foundation cool. One way is to find a similar machine on a similar duty, and to use an electronic contact thermometer to plot a temperature profile, from foundation to bearing. The effective temperature may be considerably influenced by the type of machine support (e.g., centre-line, wobble-foot, or integral), by any water cooling of the bearing pedestal, by thermal insulation of the machine casing or its supports, by wind, etc. If more specific data is unavailable, then use temperatures of 65°C for motors with the bearings in the end-shields, 25°C for pedestal bearings, and 65°C for high-speed gearboxes.

The internal clearance of rolling bearings, after fitting, is usually so small that changes of shaft location within the bearing are not significant to the alignment. In contrast, the clearance in a 100 mm bore journal bearing is of the order of 0.1 mm, and therefore change of shaft location within larger journal bearings may need to be considered. The shaft location is determined by the direction of the resultant of the radial forces acting on the shaft, and the thickness of the oil-film. The direction of the resultant can change radically; with horizontal gears, for example, the tooth reaction commonly lifts one of the shafts from the bottom to the top of its bearings during start-up (Fig. 8.2). For cylindrical bearings of normal design, Fig. 8.1 can be used to obtain the shaft attitude angle, when bearing clearance and oil-film thickness are known. Lemon bearings operate with a much larger attitude angle, and tilting pad bearings locate the shaft almost centrally, although the shaft may well be orbiting around this point.

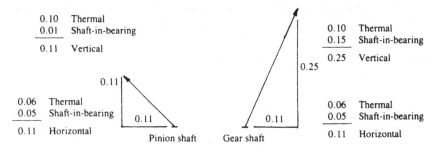

Fig 8.2   Radial shaft displacement for a typical gearbox.

Piping forces and moments, imposed on machine nozzles, are claimed to be the prime cause of alignment disturbance, even in well-engineered plants. Careful design and installation help to minimize the effect of piping on alignment; any residual effect at shutdown conditions can be observed by watching the indicator gauges as the nozzle flange bolts are loosened or tightened. But during operation, piping forces change due to thermal expansion of the piping, movement of pipe-connected equipment, etc. The flexibility of connecting pipework should therefore be considered, and increased if necessary to reduce the nozzle loads to values acceptable to the machine manufacturer. Alternatively, the common baseplate under machine and driver may be designed for sliding supports and adjustable resilient mounts. As the normal procedure is to reduce excessive piping forces to generally accepted values, their effect is not intentionally counteracted by alignment offsets.

Change in baseplate support is seldom a cause of misalignment when both machines are mounted on a common massive concrete base, although foundation settlement may increase the piping forces. The alignment of machines mounted on steelwork, however, such as offshore oil or gas production platforms, may be influenced by activities nearby, such as loading or unloading deck areas, filling or draining large storage tanks, or operating a crane mounted on the same module. As alignment offsets cannot be used to compensate for such erratic movements, the skids of larger offshore machines are either mounted on heavy stiffening steelwork added to the deck, or designed for three-point mounting such that they are isolated from harmful deck movements.

Axial alignment, too, requires consideration. Thermal expansion of the shafts causes axial displacement of the shaft-end at the coupling, and this can be considerable if the thrust bearing is remote from the coupling and the shaft is long. However, the displacement is reduced by the expansion of the machine casing, if this is fixed at the coupling end, and free to move at the other (Fig. 8.3). When membrane or diaphragm couplings are used, such displacements may cause significant loads on both couplings and thrust bearings. The alignment is then axially offset, such that the couplings are installed with a pre-stretch which reduces to zero at operating conditions. When gear couplings are used, it is not possible to reduce axial forces in this way. Large motors and generators often have shafts axially located by the coupling rather than by the thrust bearing. While this allows a large axial float, a proper axial alignment is, nevertheless, required because of the magnitude of the magnetic centering force acting on the

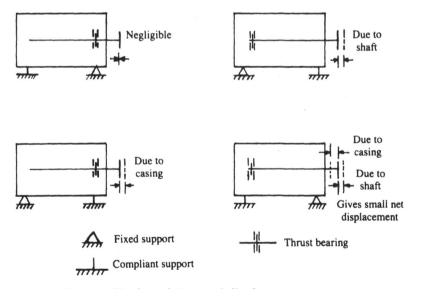

Fig 8.3 Shaft-end thermal displacements.

shaft; 5 kN, for example, for a 10 MW motor. If more specific figures are not available for the temperature of motor and generator shafts, then 80°C may be used for shafts in contact with rotor laminations, and 10°C above ambient when the poles are mounted on a ventilated spider.

Some multistage centrifugal pumps are fitted with balancing discs to reduce the hydraulic load on the thrust bearing and, with the pressure of the pumped fluid acting on them, they locate the shaft axially in much the same way as a thrust bearing. But there is a difference sometimes relevant to alignment; the rate of axial wear can be quite large if the pumped fluid is abrasive.

Note, too, that offsets for elastomeric couplings with rubber block inserts may require special consideration. When driven by electric motors, having characteristically high starting torques, such couplings may perform in a relatively rigid manner. For these couplings it is suggested that cold parallel misalignment should not exceed 0.02 mm; and that the change in alignment, from cold to hot operating conditions, should be limited to 1/500 times the spacer length, or 0.20 mm whichever is smaller. These constraints may require the machine to be mounted in some way to reduce the change in alignment, e.g., using water-cooled or centre-line supports.

Having then decided that an offset radial or axial alignment may be required, the thermal expansions are calculated, and the shaft-within-bearing locations estimated. Radial shaft displacements for a typical gearbox are shown in Fig. 8.2, and axial displacements for a typical centrifugal compressor in Fig. 8.3.

8.2 THE ALIGNMENT MAP

The alignment map is a graphical specification of the desired cold alignment, a convenient measurement work-sheet, and a record of the actual alignment achieved. It is called a map because it shows where you are, where you wish to be, and how to get there. It relates the predicted shaft displacements to the alignment during operation, instructs those adjusting the alignment, indicates the required thickness of adjusting shims, records the alignment achieved, and is the basis for any future correction which may be indicated during operation. For a complete map the plan and elevation are shown separately but, as the two are identical in principle, only the elevation is shown in Fig. 8.4.

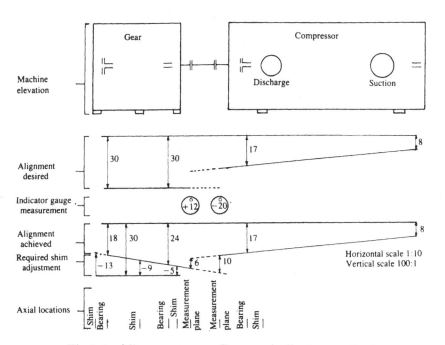

Fig 8.4 Alignment map. Reverse indicator method.

There are three basic steps when making an alignment map.

(1) In the X direction, use a scale of 1:10 to plot the locations of radial and thrust bearings, measurement planes and shimmed machine supports. In the Y direction, use a scale of 100:1 to plot the desired operating and cold locations of the shafts. The desired operating locations are, of course, such that the coupled shaft-ends have a common axis, and the desired cold locations are derived from the alignment changes calculated according to Section 8.1.

(2) Measure the alignment, and enter the indicator gauge readings on the map. Interpret the readings, and plot the actual shaft locations (for measurement and interpretation see Section 8.3). Decide whether or not the alignment should be improved (refer Section 8.4). If so . . .

(3) At each support, read off from the Y axis the required change in shim thickness, and adjust the alignment accordingly. Repeat step (2).

As described in Section 8.3, the use of a hand-held electronic alignment calculator can entirely eliminate the tedious and time-consuming effort of making an alignment map. For a complex machine train, however, the map remains an excellent method of recording, interpreting, and understanding the alignment measurements.

8.3   MEASUREMENT AND ADJUSTMENT

To align one machine to another, such that the shaft-ends are co-linear during operation, it is usual to measure the offsets of one shaft axis with respect to the other. Sometimes the levels of bearing housings, or even machine casings, are measured relative to each other, or to an arbitrary reference point; but these methods require optical telescopes or laser equipment, and specialized methods.

Indicator gauges are the usual tools for measuring alignment, both radial and axial. Some coupling manufacturers will supply them complete with brackets to fit the shafts or their particular couplings, but site-fabricated brackets and standard workshop indicator gauges are equally effective. A well-defined procedure for their use is essential; without this, measured values may be incorrectly used to deduce the alignment achieved.

There are two well-known indicator gauge methods, the traditional face and rim method, and the newer reverse indicator method. Both require an indicator mounted on one shaft or coupling hub, with the plunger resting on the other. Figure 8.5 illustrates the methods. Of the two, the latter is generally preferred because the readings so obtained are independent of shaft axial float. Where the indicator span exceeds the coupling diameter, as is usual with spacer couplings, it is also a more sensitive measurement. However, the face and rim method is also acceptable provided that care is taken to prevent axial float of the shafts from causing spurious face readings; this is seldom a difficulty with shafts in rolling bearings.

Whichever method is used, dismantling of the coupling should be avoided if at all possible. There are three reasons for this; it eliminates errors due to coupling run-out (since the shafts can only be turned together), it reduces alignment man-hours, and it eliminates the slight but cumulative and almost inevitable damage to the precision surfaces of the coupling and its oil seals and fasteners. The ability to align without dismantling is an attribute which should be considered during coupling selection.

The indicator readings are used to determine the angles and offsets of the shaft axes relative to each other, and hence the required adjustment.

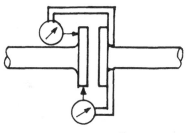

Face and rim

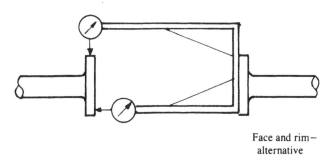

Face and rim −
alternative

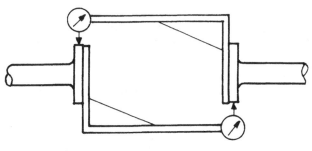

Reverse indicator

Fig 8.5 Methods of measurement.

Face readings are simple to interpret, but those less familiar may find Fig. 8.6 helpful for the radial readings. The offsets are best recorded graphically, in plan and elevation, on an alignment map – see Fig. 8.4.

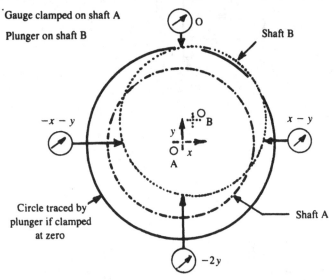

Fig 8.6   Measurement of shaft offsets using indicator gauge.
Notes.
(1) The sum of the horizontal readings is equal to the sum of the vertical readings
(2) The difference between opposite readings is twice the shaft offset
(3) The direction of the offset of the shaft on which the plunger rests is in the direction of the algebraically higher reading
(4) For parallel alignment (also angular misalignment with non-spacer couplings) changing the gauge from one shaft to the other changes only the signs of the readings.
(5) For angular alignment with spacer couplings, changing the gauge from one shaft to the other changes the magnitudes of the readings.

A third and little known indicator gauge method is the Forcados technique. This measures the angle of each shaft relative to the spacer, rather than shaft relative to shaft. Because the indicator bracket spans only across the articulation, it is robust and virtually without sag, and therefore suits any spacer length, even a long intermediate shaft. This technique is particularly suited to installations where a firewall separates the driver from the driven machine, which may make the usual indicator gauge methods impossible. The Forcados technique is illustrated in Fig. 8.7.

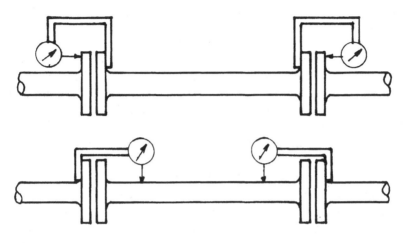

Fig 8.7   Forcados technique: alternative methods.

Indicator brackets do, of course, sag under their own weight and that of the indicator, and this introduces errors. They are, therefore, made rather stiff and, when spanning across short shaft-end separations, the error is usually ignored. But the sag of long-span brackets is easily measured; for these the readings should be corrected accordingly. Figure 8.8 shows how to do it.

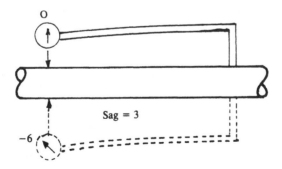

Fig 8.8   Measurement of sag.

A laser-optic development of the indicator gauge methods has become available in recent years. A laser beam from an emitter fitted on one shaft is reflected, from a prism on the other, back to a detector on the first. The detector is connected by cable to an instrument resembling a desk-top calculator. Dimensions from the point of articulation to the holding-down bolts are keyed-in; rotation of the shafts then provides all required data, without the need for taking readings or recording them. Finally, the shim adjustments are automatically calculated and displayed. The method appears particularly suited to machines separated by long spacers, and for aligning a large number of machines in a short time, e.g., during plant commissioning. The instrument eliminates errors in reading, recording and interpreting measurements, and may, therefore, be used by persons less experienced in alignment.

Similar, but simpler and cheaper hand-held electronic alignment calculators are available without the laser and electronic measurement features, some of these being intrinsically safe for use in hazardous areas. Dial gauge readings are entered at the keyboard, together with coupling-to-bolt dimensions, after which the instrument displays the shim adjustments required. For the less experienced, step-by-step instructions are also displayed. Such calculators eliminate the need for alignment maps, and hence make irrelevant the main disadvantage of the reverse indicator method – that the required shim adjustments are not so easily determined as with the face and rim method.

Whatever measurement method is used, the measurement condition should be recorded; for example ambient temperature 10°C, standby temperature, operating temperature, or hot alignment; the significance of these different conditions was explained in Section 8.1.

Unfortunately, all the foregoing methods have a basic limitation; they cannot be applied to running machines. And as shaft movement, from measurement condition to operating condition, is seldom known with certainty, the operating alignment is also in doubt. For this reason a hot alignment check is often advocated; that is a check measurement as soon as possible after shut-down and dismantling of coupling guard, etc. But, because dismantling may take quite some time, and because the rate of cooling is greatest immediately after shut-down, the value of a hot alignment is often questionable.

Bearing housing movements can, however, be measured during operation using the Essinger method. Four steel balls are permanently set in a vertical plane near each bearing, two on the bearing housing, and two on the machine foundation. An inclinometer and inside micrometer are used to measure angles and expansions, from which the vertical and horizontal movements of the bearing housing can be deduced.

The basis for adjusting the alignment is the calculator display or the alignment map (Fig. 8.4), which shows the shim thicknesses to be used, and the horizontal movements required. Vertical alignment is corrected first, by adding or removing shims at the holding-down bolts. Shims of stainless steel are preferred for maintaining precise alignment over many years. To minimize their number, a range of thicknesses is used, typically 0.05, 0.10, 0.20 and 0.50 mm, and these are complemented as necessary with stainless steel plates of thickness 1, 2, 5 and 10 mm. Pre-cut peelable shim-packs from specialist suppliers are a convenience, as are specially-designed jacking chocks, supported on poured epoxy grout.

The horizontal alignment of small machines may be adjusted by blows from a soft hammer, but for controlled, precise movement of larger machines, permanently fitted horizontal jacking screws are used. The total process of measurement, adjustment, and check measurement is likely, in practice, to be iterative, although precision at each stage reduces the number of iterations, and the man-hours required.

8.4 ALIGNMENT LIMITS

It can be important to have some idea of how much misalignment of shafts is tolerable, not only because exceeding the allowable limits may risk an unscheduled shutdown, but also because the limits indicate the accuracy required when setting the alignment. Precision setting of alignment is time-consuming and costly, and unnecessary accuracy should be avoided. The limits, within which the measured alignment should lie, should be narrow enough to reduce machine failures, but wide enough to be set quickly and easily.

There are three independent limits to acceptable alignment. The misalignment should not exceed the capability of the coupling, it should not overload the bearings, and it should not cause instability of a rotor supported in hydrodynamic bearings. The first of these limits is normally stated in the coupling manufacturers' literature. The other two limits are usually ignored, unless bearing failure or high vibration indicates a problem. The following paragraphs therefore explain various ways of stating a coupling's capability, they suggest how bearing load and instability problems can be avoided at the time of choosing a coupling, and they indicate how to change an existing alignment if such a problem nevertheless occurs. Manufacturers define the misalignment capability of spacer couplings in terms of allowable angular and axial displacements; for non-spacer couplings radial displacement is defined as well. The angular displacement refers to each end of a spacer coupling; in contrast, the axial displacement usually refers to the sum of the displacements at each end.

For gear couplings, the coupling alignment capability limits are as explained in Section 3.2.2. The allowable angular displacement is inversely proportional to speed, and is limited by tooth wear; axial displacement is limited by the length of the teeth. A typical minimum axial displacement is 6 mm, as required in API Standard 671, although coupling axial capability need never be a constraint when using gear couplings. In contrast to diaphragm and membrane couplings, the ability of gear couplings to cope with angular displacement is quite independent of their ability to cope with axial displacement. The necessary minimum angular displacement of 0.00075 radians, to maintain adequate lubrication, seems worth reiterating here.

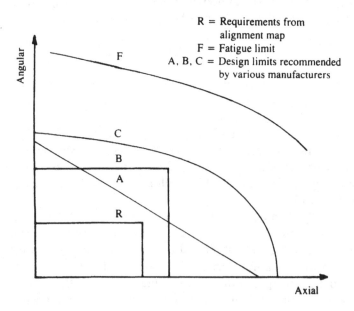

Fig 8.9 Coupling alignment limits for diaphragm couplings.

For diaphragm and membrane couplings, the coupling alignment limits are set by the requirement to avoid fatigue failure of diaphragm or membranes. Axial displacement, speed, and torque all result in steady stress, but angular displacement causes a cyclic fatigue stress. Using the infinite life criterion, the fatigue failure limit for a particular coupling may be drawn on a graph of angular versus axial displacement (Fig. 8.9), thus defining all theoretically acceptable combinations of the two displacements. For a particular application, if the design limits recommended by various coupling manufacturers are added to such a diagram, the limits may appear as a curve, or a straight line, or may even show that the two allowable displacements are apparently independent of each other! Such diversity is perplexing. However, because there is a large factor of safety between the design limits and the fatigue limit, these diverse forms of the design limits are of little practical significance. For those selecting couplings, the important point is to obtain the coupling manufacturer's design limits, and to ensure that they are not exceeded. It can be helpful to note the angular and axial misalignment requirements from the alignment map, and to add these to the diagram as shown. A diagram, comparing contoured diaphragm with multi-membrane and corrugated membrane couplings, is likely to show that the diaphragm coupling will have the lowest limit, and the corrugated membrane coupling the highest, i.e., the latter is the most flexible. Note that the limit for multi-membrane couplings rises with a decreasing number of bolts; changing from eight to six, for example, raises the limits by half. It is also worth noting that, although the length of the coupling spacer does not affect the coupling alignment limit, a longer spacer does allow more misalignment between machines. API Standard 671 requires a spacer of minimum length about 460 mm, and this is adequate for most machines.

Misalignment and bearing load are interrelated, as explained in the chapters on each type of coupling. Thus, by selecting some arbitrary but plausible value of bearing load due to misalignment (e.g., 20 percent of bearing load due to rotor weight), a corresponding misalignment can be determined; and this can be drawn on an alignment map to define the alignment limit for bearing load. In practice, corrugated membrane, multi-membrane, and contoured diaphragm couplings are generally so soft that normal misalignments do not significantly affect bearing loads. In contrast, it is suggested that alignment limits for bearing loads should be determined when using a quill shaft or a gear coupling. It may seem paradoxical, but a gear coupling can cause bearing loads as great as those due to the weight of the rotor; the radial direction of these loads, with respect to the direction of the misalignment was explained in Section 3.4.2.

Light-weight rotors running in hydrodynamic bearings may be either stabilized or destabilized by misalignment, depending upon its direction. If the misalignment is such that a significant upward force on the shaft results, then the shaft will run more nearly concentric with the bearing and, being thereby destabilised, the increased vibration may trip the machine. Using the same plausible value of bearing load due to misalignment as in the last paragraph, an alignment limit for stability can also be drawn on the alignment map. But note that, while these two alignment limits for bearing load and rotor stability are the same in magnitude, their directions differ. However, it is usual to ignore the stability limit unless a high value of vibration suggests that a problem exists. If then X and Y proximity transducers indicate that the shaft eccentricity ratio is too small, the magnitude of the alignment adjustment, required to increase the bearing load by some significant amount can readily be determined. For gear couplings, the direction of the required adjustment, relative to the shaft attitude angle, is shown in Fig. 3.7, for other couplings, the direction of the adjustment is easy to visualise.

CHAPTER 9

Effect on Machine Train Dynamics

9.1 TORSIONAL DYNAMIC RESPONSE

9.1.1 Introduction

To enhance system reliability it is often prudent to carry out a torsional vibration analysis. In many cases these calculations are done by the machine vendor or by the coupling manufacturer on his behalf.

Many purchasers require such an analysis for unspared single stream machines, for machines subject to torsional excitation, and for machines sensitive to the effects of torsional vibration. To be useful, the report on the analysis should include all the input data and assumptions, the type of analysis (natural frequency or forced damped), and the computer program reference. It should conclude with a clear statement that the design or system is acceptable, or that a certain speed range should be avoided, or it should propose specific modifications to the system.

Excessive torsional vibration may result in tooth failure of gears and gear couplings, worn or overheated elastomeric coupling elements, fretting of keys, keyways and the bores of couplings and impellers, and the fracture of shafts.

Specifically, torsional analysis should be considered for systems including the following:

Reciprocating machines;
Centrifugal pumps, fans, compressors and agitators;
Synchronous motors;
Gears;
Gear Couplings.

Torsional analysis is, in the main, carried out by the trained specialist analyst. Nevertheless, it can be helpful if design and plant engineers have some understanding of the concept.

A torsional analysis consists, in general, of two distinct calculations.

(a) The calculation of natural frequencies and mode shapes.
(b) The calculation of shaft angular displacement amplitudes and stresses, elastomeric coupling heat loads, and vibratory torque values.

Where systems include an induction motor, synchronous motor, or generator, a transient analysis may be carried out to determine the acceleration and dynamic torques during start-up. If calculations indicate that the system is not satisfactory and, therefore, requires modification, then options include the following.

Additional damping:
- change to an elastomeric coupling;
- change rubber material or hardness of the elastomeric coupling;
- add a vibration damper.

Increased mass:
- add a flywheel.

Redistribution of mass:
- change driving flange and outer coupling (elastomeric type) from driving to driven shaft, or vice versa.

Change of stiffness:
- modify shafting or coupling spacer;
- change coupling size or type;
- change rubber block shape, material or hardness.

9.1.2 System model

To simulate the system response it is first necessary to produce a mass elastic model. This simply represents each part of the system as an equivalent stiffness and an equivalent mass. Figure 9.1 illustrates such a system for an electric motor and reciprocating compressor where 1 and 2 are the cylinder inertias, 3 is the combined inertia for compressor flywheel and the outer half of an elastomeric coupling, 4 is the coupling inner member inertia, and 5 is the motor shaft inertia.

The equivalent stiffness values have been calculated from the manufacturer's shaft drawings and the coupling stiffness has been obtained from published dynamic data.

9.1.3 Natural frequencies

It is quite possible to calculate the natural frequencies for simple systems using the standard formula $F = (1/2\pi)\sqrt{\{K(\mathcal{J}_1 + \mathcal{J}_2)/(\mathcal{J}\mathcal{J}_2)\}}$ Hz or by using the well known Holzer method. Today, however, most analysis is carried out using matrix computer programs which can cope with a large number of masses, and complex gearing and multiple prime movers.

For the matrix solution, the equation of motion including damping is expressed in matrix notation as

$$[\mathbf{K}_x]\theta_a + [\mathbf{C}_x]\dot{\theta}_a = [\mathbf{M}_x]\ddot{\theta}_a$$

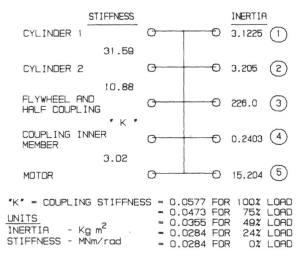

Fig 9.1 **Mass elastic system.**

Where

M_x = mass matrix
C_x = damping matrix
K_x = stiffness matrix
$\ddot{\theta}_a$ = acceleration vector
$\dot{\theta}_a$ = velocity vector
θ_a = displacement vector

The damping matrix $[C_x]$ is normally small, and in torsional systems has little effect on the natural frequencies. Therefore, the damping matrix $[C_x]$ can be assumed to be 0 when the torsional natural frequencies and mode shapes are calculated.

The stiffness and mass matrix are therefore related in the equation

$$[K_x]\theta_a = [M_x]\ddot{\theta}_a = -\omega^2[M_x]\theta$$

Where ω is the frequency of vibration (Hz)

The Eigen values and Eigen vectors of the matrix are then calculated to obtain the natural frequencies (Eigen values) and the modal shapes (Eigen vectors) of the system. Figure 9.2 shows how a six-mass system is handled.

Couplings and Shaft Alignment

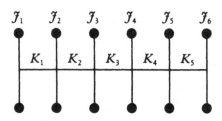

$$
\begin{bmatrix}
K_1 & -K_1 \\
-K_1 & K_1+K_2 & -K_2 \\
0 & -K_2 & K_2+K_3 & -K_3 \\
0 & 0 & -K_3 & K_3+K_4 & -K_4 & 0 \\
0 & 0 & 0 & -K_4 & K_4+K_5 & -K_5 \\
0 & 0 & 0 & 0 & -K_5 & K_5
\end{bmatrix}
\begin{bmatrix}
\theta_{a1} \\ \theta_{a2} \\ \theta_{a3} \\ \theta_{a4} \\ \theta_{a5} \\ \theta_{a6}
\end{bmatrix}
= -\omega^2
\begin{bmatrix}
\mathcal{J}_1 & 0 & 0 & 0 & 0 & 0 \\
0 & \mathcal{J}_2 & 0 & 0 & 0 & 0 \\
0 & 0 & \mathcal{J}_3 & 0 & 0 & 0 \\
0 & 0 & 0 & \mathcal{J}_4 & 0 & 0 \\
0 & 0 & 0 & 0 & \mathcal{J}_5 & 0 \\
0 & 0 & 0 & 0 & 0 & \mathcal{J}_6
\end{bmatrix}
\begin{bmatrix}
\theta_{a1} \\ \theta_{a2} \\ \theta_{a3} \\ \theta_{a4} \\ \theta_{a5} \\ \theta_{a6}
\end{bmatrix}
$$

Fig 9.2 Matrix for six mass system.

The calculated frequencies can be conveniently illustrated on a Campbell diagram. Figure 9.3 shows such a diagram for an engine–gearbox–compressor system (Fig. 9.4). The diagram shows frequency against speed for the various order numbers and clearly identifies the various criticals — that is, coincidences of natural and exciting frequencies in the speed range. In the case of the one-node mode (or first mode) it was calculated at 18 Hz determined mainly by the coupling. The three-order is of major importance; it has been calculated at 307 r/min (5.1 Hz) engine speed and is, therefore, well below the speed range.

The two-node mode was calculated at 74 Hz, and is a resonance of the gearing between the two couplings, with very little excitation of the rest of the system.

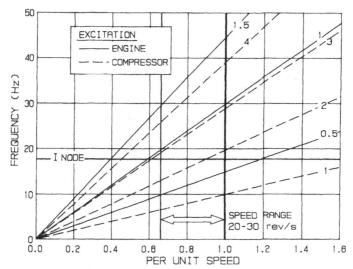

Fig 9.3 Campbell diagram.

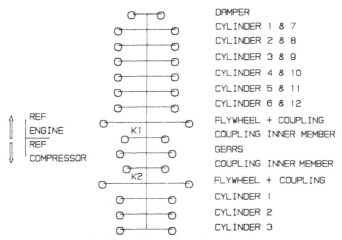

Fig 9.4 Mass elastic system.

The three-node mode calculated at 109 Hz is a characteristic engine frequency with the magnitude of the vibration displacement being controlled by the engine damper.

The diagram shows the first mode superimposed over the engine and compressor excitation, and confirms that the stiffness and damping values of the chosen couplings are satisfactory.

The treatment of this system requires special care as the harmonics from the engine are not coincident with those of the compressor, making two calculations necessary.

9.1.4 Excitation torques

For reciprocating engines, the excitation torques are a function of the type of engine (two or four stroke, diesel or petrol) and the mean effective pressure. Harmonic coefficients have been plotted in various text books, but with the significant increase in mean effective pressures in the last few years, a fourier analysis of an actual cylinder pressure diagram yields more accurate results.

The torsional excitations generated by reciprocating compressors are determined by the bore, stroke, and the properties of the gas.

Either the excitation torques and phase angles for each harmonic will be provided by the compressor manufacturer, or a fourier analysis can be performed if a torque/crank angle diagram is available.

The same data as used for the natural frequency calculation is used to carry out the forced damped calculation. At each increment of speed over a chosen speed range, the program analyses the system response due to the excitation for each harmonic up to the 12th order. Whilst more than 12 orders can be calculated, this number is normally adequate, and in many cases some computing time can be saved by specifying a lesser number, although care should be taken not to miss a significant order. The program then phases each harmonic together to produce a synthesis for each value of speed. Results can be displayed graphically in the form of speed against value of torque, stress, amplitude, or coupling power absorbed, at any point in the mass elastic system. The values obtained are then compared to allowable values. Manufacturers have their own standards for these, but suitable criteria can also be found in the API standards 617 and 618, US MIL. standard 617, and publications by Lloyds Register of Shipping.

Figures 9.5, 9.6 and 9.7 illustrate the output from this type of analysis, using the system illustrated in 9.1. In this case the only important parameter is the coupling-absorbed power, as other parameters were found to be well within acceptable limits.

The graphs illustrate 24 percent load, 49 percent load, and 100 percent load respectively, using a 12th order synthesis with the coupling manufacturer's allowable value superimposed. In each case, the curves show that there are no significant criticals on or near the running speed. The flanks of the criticals at the running speed give vibratory levels in the system well within the allowable parameters.

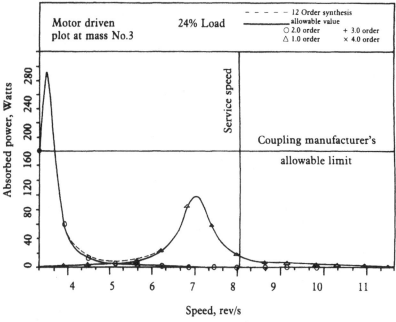

Fig 9.5

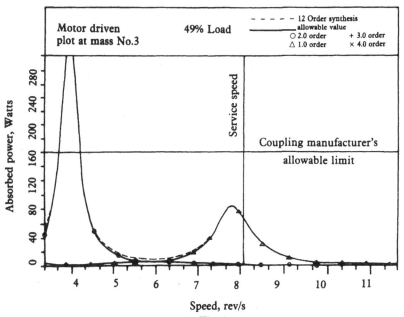

Fig 9.6

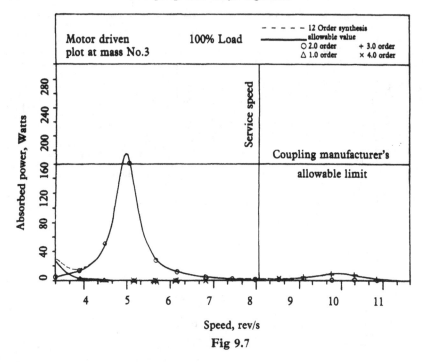

Speed, rev/s

Fig 9.7

9.1.5 Transient analysis

From experience with synchronous motors, excitation from the motor can produce a significant response in the first torsional mode of vibration only. Again, a mathematical model of the drive system, incorporating the inertia and stiffnesses of the rotating components, is idealized from the data available (Fig. 9.8). In this case the system need only be based on a two-mass model, as only the first torsional mode will need to be examined.

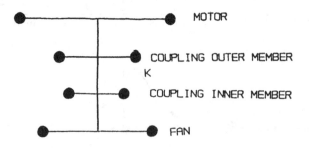

Fig 9.8 Mass elastic system.

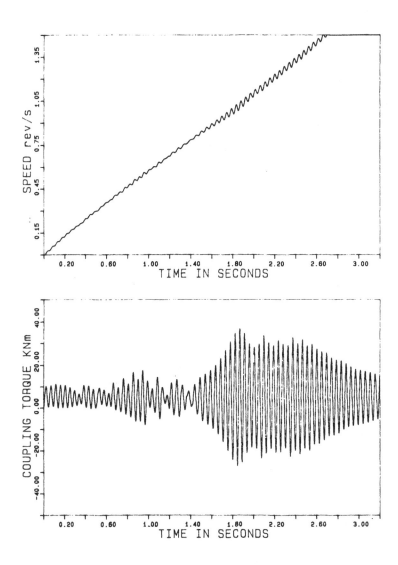

Fig 9.9 Transient analysis (solid coupling; fan drive).

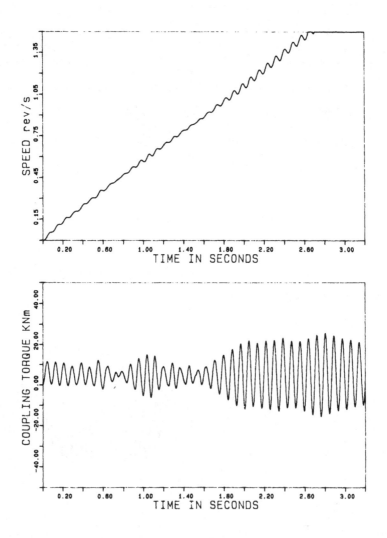

Fig 9.10   Transient analysis (non-elastomeric coupling; fan drive).

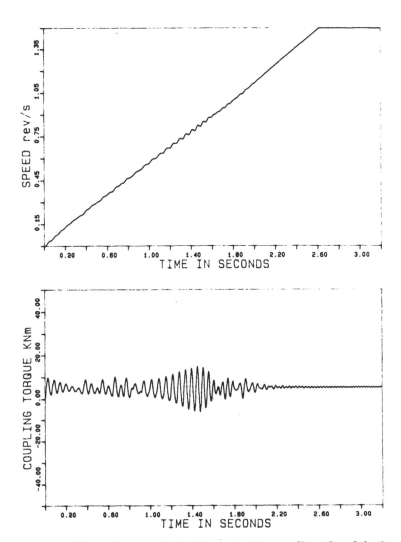

Fig 9.11   Transient analysis (elastomeric coupling; fan drive).

Using excitation data from the motor manufacturer, computer calculations are carried out using Runge–Kutta numerical methods. For each system analysed, accelerations and dynamic torque are calculated during start-up; Figs. 9.9, 9.10, and 9.11 illustrate the calculation systems with a solid coupling, a non-elastomeric coupling, and an elastomeric coupling, respectively.

9.1.6   Simplified torsional solution

Although an accurate analysis requires a sophisticated set of calculations a simple approach for examining an elastomeric coupling can sometimes be adopted, using a two-mass system.

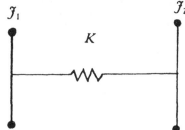

Where

$\mathcal{J}_1$ = prime mover inertia (Kgm^2)
$\mathcal{J}_2$ = driven machine inertia (Kgm^2)
K = coupling stiffness (Nm/rad)

The following calculation steps can be made.

(1)   Calculate the total inertia of the driver (e.g., motor, flywheel, outer member of coupling) $\mathcal{J}_1$.

(2)   Calculate the total inertia of the driven machine (e.g., generator and inner member of coupling) $\mathcal{J}_2$.

(3)   Select a coupling based on the factors outlined in the earlier chapters for normal torque, peak torque, and vibratory torque. Obtain a coupling stiffness, K.

(4)   Calculate the natural frequency from

$$\omega = \frac{1}{2\pi} \sqrt{ \left\{ \frac{K_{\mathrm{DYN}}(\mathcal{J}_1 + \mathcal{J}_2)}{\mathcal{J}_1 \mathcal{J}_2} \right\} } \, (\mathrm{Hz})$$

(5)   Calculate the speeds at which various orders of vibration occur. An approximation can be made where the major orders are the orders which correspond to multiples of half the number of cylinders for a four stroke engine and multiples of the number of cylinders for a two

stroke engine (e.g., for a six cylinder four stroke engine the major orders are the third, sixth, ninth).

Therefore, the speeds which correspond to the natural torsional frequency divided by the major orders will be the speeds at which significant resonance will occur.

Having chosen an order number, N_o, then calculate S, the resonant speed, from

$$S = \omega/N_o \ (r/s)$$

(6) Calculate the excitation moment for the chosen order from

$$T_{ext} = T_n \times \frac{\pi}{4} \times D_{cyl}^2 \times R_{cyl} \times N_{cyl} \times 10^{-3} \ \text{Nm}$$

Where

D_{cyl} = Cylinder diameter (mm)
R_{cyl} = Crank radius (mm)
N_{cyl} = Number of cylinders
T_n = Harmonic excitation force (N/mm^2)
for the chosen order

(7) Calculate the maximum vibratory torque from

$$T_{max} = \frac{T_{ext} \times M_{DYN}}{\{1 + (\mathcal{J}_1/\mathcal{J}_2)\}} \ (\text{Nm})$$

Where M_{DYN} is the coupling dynamic magnifier based on the chosen rubber material and hardness value outlined in Chapter 6.

This value should not exceed the maximum torque value of the chosen coupling.

(8) Calculate the vibratory torque T_s at the service speed, N_s, from

$$T_s = \frac{1}{1 - (N_s/S)^2} \times \frac{T_{max}}{M_{DYN}} (\text{Nm})$$

This must be corrected to 10 Hz by

$$T_{vib} = T_s \sqrt{\left\{\frac{N_o \times N_s}{600}\right\}} \ (\text{Nm})$$

This value should not exceed the vibratory torque published for the chosen coupling.

Should the calculated vibratory torque or the calculated peak torque be excessive then the calculation should be carried out again using the various options outlined earlier in the chapter. The first option to consider would be a change in element hardness and hence a new value of *K*. A program suitable for a programmable calculator or PC is given below.

```
10 REM - TWO MASS TORSIONAL CALCULATION TO
15 REM - ASSESS FLEXIBLE COUPLING SELECTION
20 REM *************************************
30 CLS
35 PI=3.14159
40 INPUT " ENTER ENGINE + FLYWHEEL INERTIA J1   -   ",ENGJ1
50 INPUT " ENTER DRIVEN MACHINE INERTIA    J2   -   ",DRIVJ2
60 INPUT " FLEXIBLE COUPLING STIFFNESS      K   -   ",STIFFK
65 rem
70 FREQ=9.55*SQR(STIFFK*(ENGJ1+DRIVJ2)/(ENGJ1*DRIVJ2))
80 PRINT:PRINT
90 PRINT " NATURAL FREQUENCY OF SYSTEM         = ";FREQ;" VPM"
100 PRINT:PRINT
110 INPUT " TYPE ORDER No. OF MAJOR CRITICAL   -   ",ORD
120 SPEED=FREQ/ORD
130 PRINT:PRINT
140 PRINT " RESONANT SPEED OF ";ORD; " ORDER   = ";SPEED;" RPM"
150 PRINT:PRINT " PLEASE TYPE IN THE FOLLOWING PARAMETERS "
155 rem
160 INPUT " BORE        = ";BORE
170 INPUT " STROKE      = ";STROKE
180 INPUT " No. OF CYLS = ";NCYLS
190 INPUT " Tn VALUE    = ";TN
200 INPUT " MAGNIFIER   = ";MAG
215 rem
220 EXTORQ=BORE*BORE*PI/4*STROKE*TN*NCYLS*0.5E-3
230 PEAKTQ=EXTORQ*MAG/(1+ENGJ1/DRIVJ2)
240 PRINT:PRINT
250 PRINT " MAXIMUM VIBRATORY TORQUE = ";PEAKTQ
260 PRINT:PRINT
270 INPUT " SERVICE SPEED = ";SRPM
280 PRINT:PRINT
290 rem
300 FLANKTQ=1/ABS(1-SRPM/SPEED*SRPM/SPEED)*PEAKTQ/MAG
310 PRINT:PRINT
320 PRINT "FLANK VIBRATORY TORQUE = ";FLANKTQ
330 PRINT:PRINT
340 REM
350 FREQCOR=FLANKTQ*SQR(ORD*SRPM/600)
355 PRINT " FLANK VIB TORQUE CORRECTED TO 10 Hz = ";FREQCOR
360 PRINT:PRINT
560 END
```

9.2 ROTOR LATERAL DYNAMIC RESPONSE

There are three ways in which couplings influence the lateral vibration of the coupled rotors, affecting critical speed and vibration magnitude. First, the mass of the coupling, and the distance of the hub from the adjacent bearings, both lower the lateral critical speeds. Second, the stiffness of the coupling can raise or lower critical speeds when shafts are misaligned. And third, coupling unbalance influences the magnitude of the resulting vibration; balancing standards are discussed in Section 9.3.

Many standard machine-coupling-driver combinations are known to perform well, and hence critical speeds need be of no particular concern to the user. However, it not infrequently happens that such standard machines are modified for some good reason, either by the manufacturer or by the user; the coupling type is changed for maintenance convenience perhaps, or a normally fixed-speed machine is changed to variable-speed. In such circumstances the lateral critical speeds of both driver and driven machines should be considered, as of course for all special-purpose machines. The calculation of rotor lateral critical speeds is described in many texts, and has been refined in a number of computer programs. Should calculation appear useful, the user can have it done by the machine manufacturer, or by independent consultants with experience in rotor dynamics. However, it sometimes happens that the dynamics of the rotor are considered without reference to the particular coupling to be used, which may be selected by others. Machine users should, therefore, be aware of the effects which couplings can have on the smooth running of their machines.

Both the mass of the coupling (more precisely the mass of one hub plus half the mass of the spacer), and the overhung moment (i.e., the sum of the products of the two masses, and the distances of their points of action from the adjacent bearing) modify the critical speeds of the rotor. Coupling mass and moment can lower the rotor critical speeds significantly, especially the second, and may bring them undesirably close to the operating speed range. Consideration of the lateral vibrations of coupled rotors is complex; the effect of the vibration of one rotor on the other depends on the characteristics of the coupling, e.g., weight and stiffness. However, when normal flexible couplings are used, it is usually assumed that the rotors vibrate independently of each other. With this simplification, it is easy to visualise that any of the following changes lower the critical speed:

increased hub or spacer weight;
increased distance from bearing of hub centre of gravity;
increased distance from bearing of point of articulation;
decreased diameter of shaft overhang.

To visualise the magnitude of the effects of coupling mass, coupling over-hang and reduced shaft diameter, refer to Fig. 9.12.

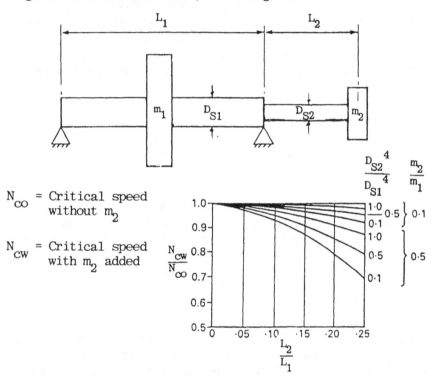

Fig 9.12 Effect of coupling on critical speed.

Computer programs used for calculating natural frequencies should take all these points into account, but there remains the important question of an appropriate margin between critical and operating speeds. Many machine standards specify minimum margins, but it should not be forgotten that some machines run successfully in spite of lateral critical speeds within the operating range; for such machines additional criteria apply. However, it may be useful to list coupling options which can be exercised to increase the critical speed when this is necessary.

- Reverse the hubs on their shafts (to bring both hub centre of gravity, and point of articulation, closer to the bearing).
- Reduce spacer weight.
- Reduce hub weight.
- Dispense with keyed fits (hydraulic fits, or integral hubs, may reduce weight).
- Use higher-strength materials (to reduce weight).
- Eliminate oil seals on adjacent bearings (normally for gear couplings only; to reduce shaft overhang).

In general, misalignment changes the critical speed of rotors running in journal bearings. Because of the coupling angular stiffness at the points of articulation, misalignment causes a change of bearing loads, which change the thickness and, therefore, the stiffness of the oil films, and hence the critical speeds of the rotors. Furthermore, since the direction of misalignment is normally unknown, the critical speeds may be either raised or lowered. Contoured diaphragm and multi-membrane couplings are relatively soft, and do not normally change bearing loads significantly; the critical speed is, in these circumstances, insensitive to misalignment. Gear couplings, however, can add loads comparable to the normal bearing loads (see Section 3.4), making the critical speeds both more sensitive to misalignment, and less predictable. When gear couplings are used with journal bearings, it therefore seems prudent to consider the sensitivity of the critical speed to misalignment to ensure that, with the worst expected misalignment, the required margin between operating and critical speeds is maintained.

It is sometimes claimed that one or other particular type of coupling is in general lighter, or has a lower overhung moment, than other types. Many machines, of course, have generous margins between operating and critical speeds, such that weight and moment are of little consequence. But for machines where margins matter, such general claims should be confirmed for the specific alternative couplings proposed.

9.3 BALANCING STANDARDS

Unbalance causes fatigue forces on bearings and their supports, and may excite resonances in piping and other nearby items. For flexible rotors (i.e. rotors running above their first lateral critical speed), unbalance also causes shaft bending, thus increasing the possibility of the rotor rubbing against the stator. The forces resulting from unbalance increase with the degree of rotor unbalance, and the square of the speed. At higher speeds they become greater than the bearing static load, unless the unbalance is appropriately limited. The amount of unbalance that can be tolerated in the coupling is determined by the dynamic response of the rotors of the coupled machines, and is therefore normally best left to the machine manufacturers to quantify, provided that machine vibration limits have been clearly specified. However, there are alternative guides to determining an appropriate permissible coupling unbalance.

The International Standards Organisation (ISO) Standard 1940, *Balance Quality of Rotating Rigid Bodies,* classifies rotor types and allots corresponding balance quality grades. The standard includes a graph such that, for any balance quality grade and rotor speed, the permissible eccentricity of the rotor centre of gravity (i.e., the permissible residual unbalance per unit of rotor mass) can be read off. These recommended balance quality grades are based on experience and, if due regard is paid to them, satisfactory running conditions can most probably be expected. The standard does not refer to couplings, but it seems reasonable to choose a balance quality grade for the coupling which equals that of the coupled rotors, such that the rotors and the coupling have a common permissible eccentricity. The permissible coupling unbalance, half taken in each of the two correction planes, is then the product of the eccentricity read from the graph, and the coupling weight. Note that ISO 1940 applies to rigid rotors, i.e., those running sufficiently below their first lateral critical speed, that shaft bending does not significantly affect the balance. This limit, for considering a rotor to be rigid, is often stated as N/N_o <0.2 where N is the operating speed, and N_c the first lateral critical speed. At ratios greater than 0.2, shaft bending into the mode shape increases the unbalance, and therefore the cyclic load on the bearings, by a factor of $[1 - (N/N_c)^2]^{-1}$. Thus, when adequate internal clearance exists, ISO 1940 for rigid rotors may be used as a guide for sub-critical flexible rotors also, and their couplings, if the permissible unbalance is multiplied by the factor $1 - (N/N_c)^2$. But as N approaches close to N_c such a criterion clearly becomes impossible to achieve. For such couplings a practicable permissible unbalance for the particular machine must be proposed, and confirmed acceptable by vibration measurement.

In spite of its origin, the scope of the American Gear Manufacturers Association (AGMA) Standard 515.02, *Balance Classification for Flexible Couplings,* is not restricted to gear type couplings. It is intended as a guide to the users of various types of high speed couplings, and is commendably simple to use. First, a graph of speed versus coupling mass is used to obtain an appropriate coupling balance selection band. For rotors having average sensitivity to coupling unbalance, a table then relates selection bands A – G, with coupling balance classes 6 – 12. For rotors having low sensitivity, the balance class drops by one, and for rotors having high sensitivity, the balance class increases by one. Finally, the AGMA balance class of a coupling is defined in terms of its maximum permissible eccentricity at the balance planes. The following factors are considered to make a rotor relatively sensitive to coupling unbalance:

long or flexible shaft extension;
high bearing load due to coupling weight
   (relative to the total bearing load);
flexible shaft support;
proximity to critical speed;
long spacer.

Knowing coupling weight and speed, it is thus possible to determine, with some judgement, but nevertheless quite readily, the AGMA coupling balance class recommended. Finally, the balance classes are defined in terms of permissible eccentricities measured in each balance plane, expressed in micro-inches. There appears to be no simple correlation with ISO 1940, although the objective of a smooth running rotor system is common to both Standards.

The American Petroleum Institute (API) Standard 671, *Special-Purpose Couplings for Refinery Services*, is a users standard applicable to couplings for equipment trains that are normally unspared, and critical to the continued operation of the refinery. Such couplings are usually engineered and manufactured specifically to meet the operating conditions of the equipment trains in which they are installed, which may operate either below or above the first lateral critical speed. The standard is widely respected in the oil, gas, and petrochemical industries worldwide, and a number of coupling manufacturers offer gear, contoured diaphragm, and multi-membrane couplings which comply. Such a high standard is, of course, costly to achieve, and more rigorous than most machines require. The API concept of specifying coupling balance is quite different from both ISO and AGMA. The only option in the standard is a choice of one of three balancing procedures; each procedure defines a related maximum permissible unbalance in terms of coupling weight and speed.

The three procedures and permissible eccentricities, measured at the balancing planes, are:

(1)   Component balance (1/15N)*
(2)   Component balance and coupling assembly check balance (1/1.5N)
(3)   Component balance and coupling assembly balance (1/15N)*
(*or an unbalance of 7g–mm, if greater)

Procedure (1) is such that balanced coupling components held as spares may be fitted to the coupling in operation, should this be necessary, without further balancing (i.e., balance correction) or even check balancing (i.e., determination of residual unbalance). The residual unbalance of the assembled coupling is therefore unknown.

Procedure (2) is as Procedure (1), except that the assembly is check balanced. If not within the permissible limit, then the set-up, assembly or component balance must be improved until the limit is met. The final residual unbalances are recorded. No balance corrections are allowed on the assembled coupling. If spare coupling components are fitted, check balancing must follow.

Procedure (3) is as Procedure (2) but, after satisfactory check balance within the (1/1.5N) limit, the assembly is corrected to a limit of (1/15N). No guidance is given on the choice of procedure; in practice the choice should probably be a matter for discussion between the purchaser and the manufacturers of the driver and driven machines.

It is evident from the foregoing that the specification of an appropriate permissible coupling unbalance for a particular machine is not always a clear-cut matter, partly because of the empirical nature of the vibration criteria themselves. A rigorous unbalance specification can be difficult and costly to meet, and may even be beyond the capability of available balance machines. On the other hand, a specification which is too lax will result in rough running of the machine, and increased maintenance of bearings and seals. There may be some circumstances, therefore, when experimental confirmation of the stated permissible unbalance is worthwhile.

Confirmation preferably should be obtained by spinning the rotor, complete with coupling and simulated spacer, in the machine itself. Assuming that the angular position of the coupling balance is unknown, an unbalance equal to the permissible coupling unbalance is added to the coupling at a random angular zero position, and the change in synchronous vibration noted. The unbalance is relocated successively at positions $\pi/2$, π, and $3\pi/2$; the vibration response indicates whether the permissible unbalance may be increased, or should be decreased. For flexible rotors, the possible loss of internal clearances should of course be considered before doing such a test.

In summary the following suggestions are made concerning coupling balance.

(1) Use API 671 for couplings for machines within its scope. Such couplings are required to be balanced, but the choice of balance procedure should be discussed with the manufacturers of the driver and driven machines, in order to achieve an adequate balance, and yet avoid an unnecessarily rigorous procedure.

(2) For other machines having rigid rotors ($N/N_c < 0.2$), select a balance quality grade from ISO 1940 for the type of machines concerned. The balance quality of manufacturers' standard couplings will not, in most cases, need to be improved.

(3) For sub-critical flexible rotors again use ISO 1940, but reduce the permissible unbalance by multiplying by the factor $1 - (N/N_c)^2$. Discuss with the machine manufacturers.

(4) For super-critical rotors, discuss with the machine manufacturers.

(5) When the unbalance of a coupling is measured, retain a copy of the record, and the related procedure.

(6) When the adequacy of a permissible coupling unbalance for a particular machine is in doubt, confirm it by experiment.

CHAPTER 10

Hub Mounting

For crankshafts and large high speed gears, coupling hubs and shafts are sometimes forged as one piece, thus minimizing the overhung moment on high speed shafts, and completely eliminating the possibility of fretting. Integral hubs are not normally used for centrifugal pumps or compressors because they do not permit assembly of the seals. Most shaft and coupling hubs are, therefore, fitted together using one of four different methods:

Key and keyway;
Shrink fit;
Friction lock or wedge lock;
Hydraulic fit.

10.1 KEYS AND KEYWAYS

Keyed connections are primarily suited for the transmission of torques where the loads are steady, or fluctuate very little. Their main advantage is that couplings and shafts can be assembled and dismantled quickly, without the aid of special devices. However, keyed connections always weaken the shaft and hub, producing stress concentrations at the keyway corners and increasing the possibility of fatigue failure. Heavy shock loads can lead to plastic deformation, and an alternative method of connection should be considered for such applications.

ISO/R773 and ISO/R774 should be referred to for general guidance on the following keys:

Square parallel;
Rectangular parallel;
Square taper, gib-head and plain;
Rectangular taper, gib-head and plain;

Three classes of fit meet the varying requirements:

Free – where the hub is required to slide over the key when in use;
Normal – where the key is to be inserted in the keyway with minimum fitting;
Close – where an accurate fit of key is required. API recommend a 70 percent blued fit for this condition.

A free fit is seldom preferred, but can be used on very small couplings, often with a grub screw through the coupling body onto the key to provide a measure of axial restraint. For mass production a normal fit is generally used, but for the majority of couplings a close fit should be specified, where fitting of the key will be required under maximum metal conditions. API recommend a 70 percent blued fit for keyway fits. For this type of keyway a bore tolerance of H7 with a shaft tolerance of m6 is recommended. However, it is not unusual for the user to specify his own tolerances according to the application.

When designing a keyed connection it is desirable to calculate the compressive load on the key using the key face actual contact area and the maximum required transmitted torque values. Two keys spaced at 180 degrees are sometimes required following this simple analysis, and are sometimes preferred because they simplify the balancing procedure.

10.2 SHRINK FIT

Where removal of the coupling hub is seldom required a shrink fit can be considered. This has the advantage that keyways can be eliminated, thus saving cost and avoiding any keyway stress concentrations. The torque transmitted is dependent on the coefficient of friction, the area of the mating surfaces, and the degree of interference. Where shock loads or variations in torque have to be considered then a safety factor should be used, preferably with a value between two and three.

Interference
The interference used should be small enough to facilitate assembly and disassembly, but large enough to prevent fretting due to minute oscillating movement of hub on shaft. To avoid yielding of the material, the limiting interference should be about 0.3 percent of the bore. Centrifugal effects also cannot be ignored, particularly in high speed applications (above 60 m/s), as this has a negative effect on the interference. Loss of interference due to centrifugal forces on steel hubs and shafts can be expressed as

$$\Delta L = 0.32 \, N^2 \, D^2 \times 10^{-12}$$

Where
  ΔL = Loss of interference (mm/mm)
  N = Shaft speed (rev/s)
  D = Hub diameter (mm)

Coefficient of friction
Wide variations can occur in the coefficient of friction. Many values have been published, but for steel components, heat shrunk with normal cleanliness, a value of 0.14 can be assumed.

Mating surface area
This is generally assumed to be the total length of the hub bore. The maximum permissible torque that can be transmitted is

$$T_M = \frac{\pi d^2 l P_m \mu}{2 F_T} \tag{10.1}$$

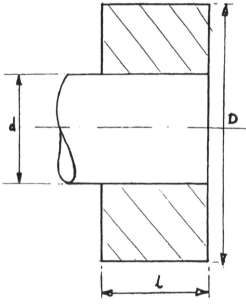

Where
T_M = maximum permissible torque (Nmm)
d = inside diameter (mm)
l = length of mating surface (mm)
P_m = pressure between mating surfaces (N/mm^2)
μ = coefficient of friction
F_T = safety factor

With the exception of the very small couplings, steel will be used for the coupling hubs, and therefore, using a solid steel shaft, the following can be shown to be valid

$$\frac{\Delta_E}{d} = \frac{P_m}{E}\left(\frac{2}{1 - (d/D)^2}\right) \tag{10.2}$$

Where

Δ_E = the effective degree of interference (mm)
E = Youngs modulus (N/mm^2)

The maximum stress in the outer components can be shown to be

$$\sigma = \frac{P_m}{1 - (d/D)^2}\sqrt{\left\{3 + (d/D)^4\right\}} \tag{10.3}$$

This should not be allowed to exceed 85 percent of the yield stress of the hub material.

From equation (10.1) the pressure between the mating surfaces can be established and substituted into (10.2) which will give the required degree of interference. Finally, from equation (10.3) the stress can be calculated which can be compared to the yield point of the hub material. The hub and shaft can then be dimensioned accordingly.

Assembly and disassembly
The assembly of the hub is accomplished through the application of heat, or by freezing, or by a combination of both. Immersion in an oil bath or heating in an oven are customary methods.

Once the interference fit has been obtained by measuring the components, the temperature increase needed to expand the coupling hub may be calculated from

$$\Delta T = \Delta e / \alpha d$$

Where

Δe = diametral expansion requirement (mm)
       (measured interference + additional allowance)
α = coefficient of linear expansion (mm/mm°C)
d = diameter of bore (mm)
ΔT = increase in temperature above ambient

To disassemble the shrink fit the axial force may be calculated from

$$F_A = \mu \pi dl P_m$$

Where $dl P_m$ are as before, but μ, the coefficient of friction, should be approximately twice the initial value to ensure that the hub puller is adequately designed.

For shafts up to about 70 mm, mechanical pull-off devices can be used, without generally, the application of heat. At least four high grade tensile bolts should be used, and the centre bolt should be provided with a fine thread (Fig. 10.1). The design of puller should prevent skewing, in order to prevent galling, and facture of the jack screws. For this reason, a puller that pushes the hub off the shaft from the back is not recommended.

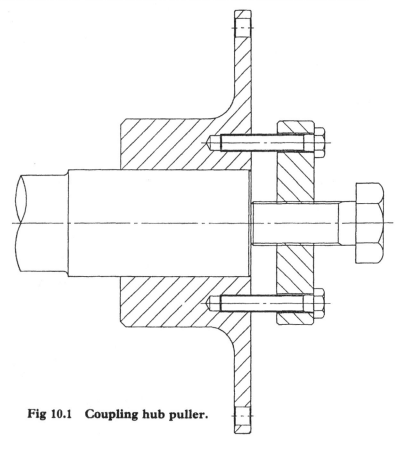

Fig 10.1 Coupling hub puller.

10.3 WEDGE LOCK

Most of these locking devices exhibit the same advantage as the shrink fit, in that neither shaft nor hub have keys. Unlike shrink fits, however, hub heating is eliminated, as is the need for extremely close machine tolerances. The elimination of heating is particularly relevant to machines requiring on-site maintenance, which are installed in an area where there is risk of explosion (e.g., a refinery). A friction lock requires less skill than a shrink fit, and removal of the hub from the shaft is quick and easy. The main disadvantage is the initial higher cost of the fitted hub.

The simplest form of these devices is the taper lock bush, which is often used to connect the smaller coupling hubs (Fig. 10.2). It requires a machined taper in the bore of the hub plus two 'half' tapped holes in the hub, with a standard key in the shaft. The bush is fitted between the hub and shaft, the tapped holes aligned, setscrews inserted and then torqued up. Sizes range from 9 mm bore to 125 mm bore for a number of proprietary hub and bush systems.

Fig 10.2   Taper lock bush.

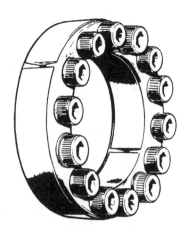

Fig 10.3   Wedge locking assembly.

Circular wedge locking assemblies (Fig. 10.3) transmit torque and axial load by radial contact pressure in the joint between the inner ring and shaft and the outer ring and hub. They are used without keys. In a shrink fit the pressure is generated by contraction or expansion, but in a wedge locking assembly the pressure is generated by radial deformation of the outer and inner rings, caused by axial displacement of the two double tapers.

When fitting it is essential to realize that the torque is transmitted by contact pressure and friction between the surfaces, hence the condition of the contact surfaces and the tightening of the locking screws are important. All contact surfaces, including screw heads and screw head bearing surfaces must be clean and lightly oiled. The hub, shaft, and locking ring should then be assembled with the locking screws lightly tightened and the hub aligned. When satisfied with the alignment, the screws should then be tightened evenly, diametrically opposite in two or three stages, until the indicated tightening torque is reached. Before the final assembly can be regarded as complete the screw torques must be re-checked.

Various proprietary designs are available, ranging in shaft size from 20mm to 1000mm internal diameter, with torques from 270 Nm to 2×10^6 Nm. If higher torque is required, and if length permits, then more than one locking assembly can be used.

10.4 HYDRAULIC FIT

The oil injection method of mounting hubs to shafts has been in widespread use for many years and provides a pressurized joint which can be assembled and disassembled without any heat or any significant amount of force. It is used, in particular, for large high speed pumps and compressors pumping flammable fluids. Oil can be introduced between the mating surfaces through oilways in the coupling hub (Fig. 10.4), connecting with a circumferential groove in the bore. Equilibrium of the parts will occur when the oil pressure equals the contact pressure between the mating surfaces, increasing the oil pressure forces the surfaces apart and oil eventually reaches the extremities of the joint. When the pressure is reduced the oil flows back through the oil supply drillings and the original contact pressure is restored.

There are four design variations; cylindrical hub bore and shaft end (without sleeve), tapered hub bore with externally tapered sleeve, tapered shaft with internally tapered sleeve, and tapered hub bore with tapered shaft end (without sleeve). Some designs incorporate 'O' ring seals, with or without backup rings, others do not.

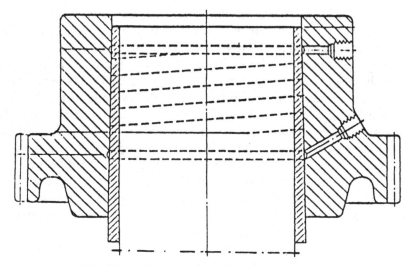

Fig 10.4   Taper hub, oil injection design.

The design calculations consist of two parts.

(a)   The taper design.
(b)   The hub and oil supply design.

For designs that do not use a taper sleeve (Fig. 10.5) then the calculation method given in Section 10.2 can be used.

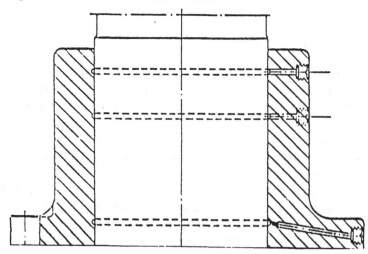

Fig 10.5   Parallel bore design.

Taper sleeve
The external taper $1/K$ should be $1/30$. Fine turning is generally sufficient and the recommended bore and shaft tolerances are:

50mm – 100mm: G6/h5
101mm – 200mm: G7/h6
201mm and above: G7/h7

d_m, the mean outside diameter of the taper sleeve, is $d_1 + l/2K$.

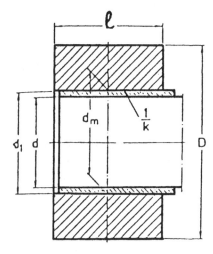

The initial contact pressure can be determined using equations (1) and (2) in Section 10.2, the effective degree of interference, ΔE, between the tapered surfaces can be obtained by substituting d_m for d.

The axial drive up distance (l_a) can be determined using

$$l_a = K (\Delta_E + \Delta_1 + 8H)$$

Where
Δ_1 = Mean tolerance on shaft diameter plus mean tolerance on bore diameter (mm)
H = Compression of surface layer
  = 0.0035

The actual degree of interference Δ_A which is obtained at maximum drive up can now be determined from the previous equation when the initial clearance is at the minimum value.

The maximum contact pressure will therefore be

$$\frac{\Delta_A}{\Delta_E} \times P_m$$

and this can be sustituted by using equation (10.3) in Section 10.2.

The mounting force F_m (N), can then be calculated from

$$= \pi \, d_m \, l \, P_m \, (1/2K + 0.02)$$

and the dismounting force, F_d (N), from

$$= \pi \, d_m \, l \, P_m \, (0.02 \; 1/2K)$$

For more detailed information, SKF publication No. P1303E (1987) "Oil Injection Methods" is recommended.

Hub design

Figure 10.6 gives a typical hub design. Oil should be forced between components at the point where contact pressure is greatest. If the section height of the outer component is more or less constant, it is best to have the groove in a central position. If contact pressure is considerable at one end of the mating surfaces, an oil groove should be incorporated in this vicinity. Dimensions of the oil distribution grooves are given in Fig. 10.7.

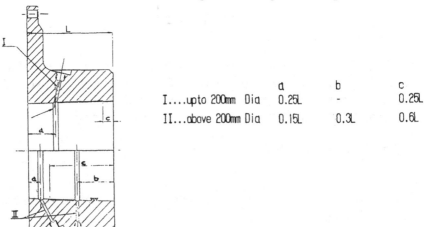

	a	b	c
I....upto 200mm Dia	0.25L	-	0.25L
II...above 200mm Dia	0.15L	0.3L	0.6L

Fig 10.6 Typical hub design.

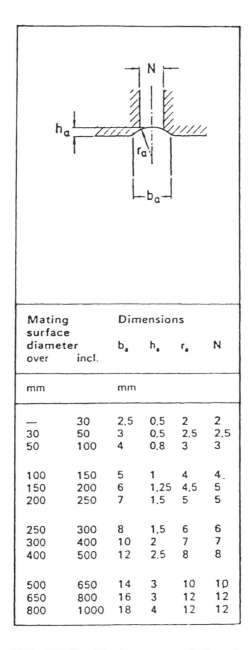

Mating surface diameter over	incl.	Dimensions b_a	h_a	r_a	N
mm		mm			
—	30	2.5	0.5	2	2
30	50	3	0.5	2.5	2.5
50	100	4	0.8	3	3
100	150	5	1	4	4
150	200	6	1.25	4.5	5
200	250	7	1.5	5	5
250	300	8	1.5	6	6
300	400	10	2	7	7
400	500	12	2.5	8	8
500	650	14	3	10	10
650	800	16	3	12	12
800	1000	18	4	12	12

Fig 10.7   Oil distribution groove design chart.

To obtain the maximum power transmitting capacity it is essential that oil should be able to drain from the mating surfaces when mounting is completed, and for this to be effective without causing delay the mounting surfaces should be provided with drainage grooves. Helical drainage grooves should normally be provided in the coupling bore; for a narrow coupling axial grooves can be used.

For hot machines, such as gas and steam turbines, the material of any 'O' rings needs consideration because the high temperature may cause early degradation of the elastomer, such that the rings can no longer retain the oil pressure during dismantling, making hub removal difficult.

Assembly and disassembly

If an hydraulic press or some other type of press is not available, then it is often necessary to provide special mounting equipment. Many different methods are available, and Fig. 10.8 illustrates a preferred type. An expanded hub has considerable potential energy, and removal, without consideration of the possibility of the hub popping off the shaft at high speed, can cause an accident. In particular, it should be noted that the oil takes time to wet the contact surfaces, and the hub may, therefore, take as long as an hour before popping off.

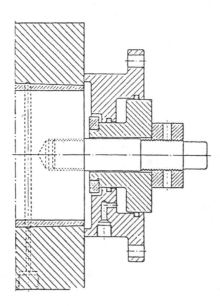

Fig 10.8   Hydraulic tool for assembly and disassembly.

Precise axial location of the hub on the shaft (or sleeve) is important as this affects the residual stresses in both hub and shaft. For this reason a two-piece stop-ring tool is sometimes specified for assembly.

Another useful tool is a plug and ring taper gauge. This is used to check both surface contact when fitting the spare coupling, and to lap hub and shaft in the event of fretting damage. When blueing of the surfaces are used, 85 percent contact area should be obtained (API requirement). It should be noted that the hub should never be lapped on the shaft, as this tends to create stepped tapers, a consequent decrease in contact area installed, and possible shaft fracture due to stress concentration at the shaft-step.

CHAPTER 11

Accessories

11.1 GUARDS

The primary purpose of a coupling guard is safety; to prevent fingers, for example, from being able to contact the moving parts. The guard may be mounted on legs on the baseplate, or flange-mounted from the machines; it may be made of solid sheet steel, or of perforated or expanded metal. Cylindrical flange-mounted guards must be axially split and, in general, must have provision for thermal expansion such that the guard does not impose a significant axial load on the coupled machines. Such provision is often made by using an 'O' ring, or a convoluted or other expansion joint. Whatever the design, the guard should be adequately robust for its environment, yet easily and quickly removable for maintenance. Some users specify that the guard shall be made from non-sparking materials, but this seems a doubtful advantage if the guard is strong enough to avoid accidental contact with the moving parts. The coupling manufacturer is not necessarily responsible for design and supply of the guard; the specification for the machine train should clearly define whether the responsibility is indeed with the coupling manufacturer, the manufacturer of either of the coupled machines, or the vendor of the machine train. Appendix C of API 671 is a useful specification, and the associated coupling data sheet is a convenient way for the user to state his particular requirements. Guards are not normally designed to retain a coupling spacer after catastrophic failure and thus prevent it from becoming an extremely dangerous missile. A better alternative is to specify a coupling designed to restrain the spacer from excessive radial movement in the event of coupling failure.

For continuously oil-lubricated gear couplings, an oil-tight guard is required, complete with spray nozzles, and oil inlet and drain connections. Some users specify windows in the guard, to observe the spray.

Windage, particularly when using large-diameter high-speed diaphragm couplings running in cylindrical flange-mounted guards, can be a problem. It may cause local sub-atmospheric pressures, thus encouraging oil to leak from the bearings into the guard. The use of minimum-moment couplings, fitted as close as possible to the bearings, exacerbates the matter. And if one of the coupled machines has its own independent oil system, as gearboxes sometimes do, even a minute leakage into the guard may eventually cause shutdown on low oil level. To prevent such leakage, a vent is sometimes mounted on the guard, either fitted with a non-rusting filter–breather or, better, piped to the oil vent system. To be effective, the vent may need to be piped within the guard to the vicinity of the bearings. Windage from these couplings can also cause very considerable heating, leading to early failure of any coupling-mounted electronics used for torque or alignment monitoring. An air purge or a lube oil spray is an effective cure. For these high-speed large-diameter diaphragm couplings, the coupling manufacturer is likely to be in the best position to provide a trouble-free guard design.

11.2 ALIGNMENT MONITORS

Since proprietary equipment has become available for on-line monitoring of shaft alignment, it has been applied to a number of major machines, with a view to increasing their reliability. There are two ways in which it can do this.

First, monitoring can detect the effect of day-to-day and long-term changes in machine support, and in piping forces, which may otherwise pass unnoticed until trouble strikes. Such changes may be caused, for example, by sunshine or rainstorms on piping, concrete shrinkage during curing, heating or cooling of foundations, or changing deck deflections on offshore platforms.

Second, monitoring verifies that the cold alignment which has been set does indeed result in an acceptable alignment during operation, both radial and axial. The ease of calculating thermal growths bears little relation to the accuracy of predicted shaft movements, and users of alignment monitors have sometimes been surprised at the discrepancies between predictions and monitor readouts.

Monitoring is perhaps more important for diaphragm and multi-membrane couplings because they give less warning of failure than do gear couplings. Diaphragm couplings, in particular, are more sensitive to axial misalignment, and monitoring can help them to achieve the infinite life for which they are designed. The monitoring of gear couplings, however, can ensure that their life is not curtailed by an alignment too perfect for effective lubrication. For verifying alignment settings, monitoring is most useful during commissioning. Where identical machines are installed, only the spare coupling spacer need be instrumented, and then fitted to each machine in turn.

An early proprietary alignment monitoring system was based on work by V.R. Dodd. It monitors the movement of one machine casing with respect to the other, by measuring the relative proximity of two overlapping bars (sometimes called Dodd bars), one projecting from the bearing housing of each machine. The monitor displays two proximity measurements in plan, and two in elevation, from which bearing movements can be deduced. Initially, the alignment must be measured with indicator gauges in the usual way, in order to provide a basis from which changes are measured. The method ignores the movements of the shafts within their bearings.

A more recent proprietary system measures the alignment of each shaft with respect to the spacer. The coupling manufacturer's standard coupling is internally fitted with electronics, and the output signal is transferred by telemetry to a non-contacting stator. Optional output signals include shaft axial location and, for gear couplings, tooth sliding velocity. The measurement is absolute, i.e., it does not rely on the measurement of changes from a basis established by other means, as does the Dodd method.

It is also possible to monitor the effects of misalignment (although not to measure the alignment itself), without using proprietary equipment; X and Y proximity transducers (to define the locations of the shafts within the bearings), and pressure gauges (to measure the oil film pressures) can be used to indicate misalignment.

11.3 TORQUE MONITORS

Unlike alignment, which is monitored to improve reliability, torque is monitored to check decline in machine efficiency, and hence to indicate the need for corrective maintenance. It is of particular interest for machines having a declining performance during operation, such as:

centrifugal compressors subject to polymerisation fouling;
steam turbines subject to deposit fouling;
gas turbines subject to compressor fouling.

It is used to measure the loss of efficiency, such that the cost of the deterioration can be calculated. Knowing the cost of a shut-down, it is then possible to determine the optimum date for on-line solvent or abrasive cleaning, or other remedial work, in order to restore the original efficiency. Torque monitoring is especially useful where two or more such machines are coupled together (e.g., a gas turbine with centrifugal compressor, or a motor with two compressor casings), because it can indicate which machine is responsible for the performance loss.

Several suppliers of torque monitors fit their equipment to standard couplings, and at least one coupling manufacturer offers his coupling complete with an optional monitor. There are two main non-contact systems. The first requires a coupling spacer fitted with strain gauges; power and output signals are transferred between spacer and stator by telemetry. With the other system, stationary electronics detect the twist occurring between toothed rings fitted on the spacer. Optional outputs from either system include speed, power, and torsional vibration.

Nomenclature

Symbol	Quantity	Units
a	Coupling mass overhang from bearing	–
b	Bearing span of machine	–
c	Clearance	mm
d	Inside diameter	–
d_1	Initial diameter of taper	–
d_m	Mean taper diameter	mm
d_t	Tooth depth	mm
e	Eccentricity	mm
f	Force on one element	N
f_m	Factor of misalignment	–
g	Acceleration due to gravity	N/m^2
h	Oil-film thickness	mm
i	Angular misalignment	Radians
j	Backlash at gear meshes	mm
l	Length of mating surface	mm
l_a	Axial drive up distance	mm
$m_{1,2}$	Masses	–
n_c	Number of cavities	–
t_c	Chordal pitch	–
w_T	Full width of teeth	mm
x	Deflection	mm
A	Area	mm^2
C	Specific damping	Nms/rad
C_x	Damping matrix	–
D_{CYL}	Cylinder diameter	mm
D	Outside diameter	mm
$D_s\, D_{s1}\, D_{s2}$	Shaft diameter	mm
D_T	PCD (Pitch Circle Diameter)	mm,m
E	Young modulus	N/mm^2
E_A	Constant	kg/mm^2
E_C	Compressive modulus	N/mm^2
E_R	Constant	kg/mm^2

F	Force	N
F_A	Axial force	N
F_d	Dismounting force	N
F_m	Mounting force	N
F_s	Service factor	–
F_T	Safety factor	–
G	Shear modulus	N/mm^2
G_A	Axial spring rate of one diaphragm assembly	kN/m
H	Compression of surface layer	mm
H_c	Cavity height	mm
H_v	Hardness (Vickers)	–
IRHD	International Rubber Hardness	Degrees
$J_{1,2}$	Inertia	kg.m^2
K	Taper	–
K_{DYN}	Dynamic torsional stiffness	Nm/rad
K_A	Axial stiffness	N/mm
K_C	Compressive stiffness	N/mm^2
K_F	Frequency stiffness correction factor	–
K_m	Load distribution factor	–
K_r	Radial stiffness	N/mm
K_s	Shear stiffness	N/mm
K_{ST}	Stiffness factor	–
K_T	Temperature stiffness correction factor	
K_x	Stiffness matrix	–
L	Spacer length	mm,m
$L_{1,2}$	Lengths	–
M	Spacer Mass	kg
M_{DYN}	Dynamic magnifier	–
M_F	Friction moment	–
M_R	Resultant moment at the coupling mass	–
M_T	Moment	–
$\mathbf{M_x}$	Mass matrix	–
N	Speed	r/s
N_c	First lateral critical speed	–
N_{co}	Lateral critical speed without m_2	–
N_{cw}	Lateral critical speed with m_2 added	–
N_{cyl}	Number of cylinders	–

N_o	Order number	–
N_s	Service Speed	r/s
P	Total force	N
P_m	Pressure between mating surfaces	N/mm²
R_{CYL}	Crank radius	mm
R	Mean element radius	–
S	Torsional resonance speed	r/s
S_c	Tooth contact stress	MN/m²
S_c'	Maximum allowable contact stress	–
T	Torque	Nm
T_{EXT}	Exitation moment	Nm
T_M	Maximum permissible transmitted torque	Nmm
T_{MAX}	Maximum vibratory torque	Nm
T_N	Harmonic exitation force	N/mm²
T_s	Vibratory torque	Nm
T_{VIB}	Vibratory torque at 10 Hz	Nm
W_c	Cavity length	mm
X	Displacement vector	–
Z	Number of teeth	–
α	Coefficient of linear expansion	mm/mm/°C
β	Angle of contact area	–
δ	Phase angle	Radians
Δ_I	Initial clearance between cylindrical surfaces	mm
ΔA	Actual degree of interference	mm
ΔE	Effective degree of interference	mm
Δe	Diametral expansion requirement	mm
ΔL	Loss of interference	mm/mm
ΔT	Increase in temperature above ambient	–
ε	Eccentricity ratio	–
Δ	Attitude angle	Degrees
φ	Tooth pressure angle	Degrees
μ	Coefficient of friction	–
σ	Stress	N/mm²
Ψ	Damping energy ratio	–
θ	Angle between direction of bearing load and direction of offset	Degrees
θ_a	Angular displacement vector	Radians

$\dot{\theta}_a$	Angular velocity vector	Radian/s
$\ddot{\theta}_a$	Angular acceleration vector	Radian/s^2
ω	Natural frequency	Hz

APPENDIX 2

Coupling Data Sheet

For couplings in accordance with Standard API 671, the use of the API 671 data sheet is most convenient because of the cross-references to the Standard.

The coupling data sheet shown on the next four pages is based on that of the API, and may be used for other couplings. It may be copied and modified by users without obtaining the permission of the Publishers. Those lines marked with a circle should normally be completed by the purchaser, and those marked with a square by the vendor. The information requested serves as an aide memoire but, depending upon the particular application and the coupling type, not all of it may be necessary. Ample space is left for purchaser and vendor to add any further data.

Coupling Data Sheet

	Doc. No.		Rev.	Page

Item No. _____ No. reqd. _____

_____

_____   Inquiry No. _____

_____   Quote No. _____

_____   P.O. No. _____

1 Applicable to ○ Proposal   ○ Purchase   ○ As Built
2
3
4
5 ○ Driver type _____ Manufacturer _____ Model _____
6   Serial No. _____   Plant Item No. _____
7 ○ Driven unit _____ Manufacturer _____ Model _____
8   Serial No. _____   Plant Item No. _____
9
10
11
12
13   **Operating Conditions**
14
15 ○ Power Transmitted (kw)   Normal _____ Maximum Continuous _____
16 ○ Speed (rpm) _____ Normal _____ Max. Cont. _____ Trip _____
17 ○ Torque (Nm)   Maximum Continuous _____ Maximum Transient _____
18 ○ Frequency of transients (Events/Time) _____
19 ○ Service factor ¦   ○ Minimum Required _____ ▢ Actual _____
20 ○ Ambient Temperature (°C) Maximum _____ Minimum _____
21   Environment   ○ Dust   ○ Hydrogen Sulfide ○ _____
22   Lubrication   ○ Continuous   ○ Filtration ▶
23   ○ Grease-Packed   ○ ISO viscosity grade _____
24   ○ Oil-Filled   ○ Pressure (barg) Normal ____ Maximum ____
25   ○ Nonlubricated   ○ Temperature (°C) Normal ____ Maximum ____
26   ▢ Flow (m³/h) Normal ____ Maximum ____
27
28 ○ For system mass-elastic data refer:
29
30 ○ For exciting vibratory torques refer:
31
32
33
34

46	
47	
48	**Coupling Data**

49	
50	□ Bolting Torque (Nm) _____ □ Lubricated □ Dry Bolt Extension ____ mm ____
51	○ Shaft Separation at Operating Temp. (mm B.S.E.) _____ Thermal Growth (mm) _____
52	From _____ ° C Ambient
53	○ Motor Rotor Float(mm) ○ Limited end Float (mm) _____ ○ Electrically Insulated
54	
55	□ Design Rating (kW/100 rpm) _____ □ Maximum Continuous Torque (Nm) ____
56	○ Required Misalignment Capability Between Shafts:
57	Steady State: Angular (Deg.) _____ Parallel Offset (mm/mm) ____ Axial (mm) ____
58	Transient: Angular (Deg.) _____ Parallel Offset (mm/mm) ____ Axial (mm) ____
59	□ Maximum Allowable Misalignment Between Shafts:
60	Steady State: Angular (Deg.) _____ Parallel Offset (mm/mm) ____ Axial (mm) ____
61	Transient: Angular (Deg.) _____ Parallel Offset (mm/mm) ____ Axial (mm) ____
62	○ Dynamic Balance ○ Component Balance only
63	○ Component Balance and Assembly Check Balance ○ Component and Assembly Balance
64	○ Check Sensitivity of Balance Machine
65	□ Maximum Allowable Residual Unbalance (g mm) Drive End _____ Driven End _____
66	□ Maximum Actual Residual Unbalance (g mm) Drive End _____ Driven End _____
67	□ Diaphragm Coupling Initial Deflection (mm) _____ □ Prestretch □ Compression
68	□ Diaphragm Coupling Axial Resonant Frequency (Hz.) _____ Calculated _____ Actual ____
69	□ Torsional Stiffness (Nm/°) _____
70	□ Axial Stiffness (N/mm) _____
71	□ Wr² (kgm²) _____
72	□ Angular (Misalignment) Stiffness (N m/°) _____
73	□ Weight on Shaft(kg) _____ Drive End_____ Driven End_____
74	□ Coupling Moment About Bearing(Nm) ____Drive End _____ Driven End_____
75	_____
76	
77	
78	**Materials**

79		Drive End	Driven End
80			
81			
82	Hub/Flange		
83			
84	Spacer		
85			
86			
87			
88			
89			
90			
91			
92	Bolts		
93			
94	Nuts		
95			
96			
97			
98			
99			

		Drive End	Driven End
103			
104	**Coupling Hub Machining**		
105			
106		Drive End	Driven End
107			
108	O Type (Integral, Cylindrical, Taper)		
109	O Taper (mm/mm)		
110			
111			
112	O Keyed or Hydraulically Fitted		
113			
114	☐ Keyway Dimensions and Number of keys		
115			
116	☐ Nominal Bore Diameter (mm)		
117			
118	☐ Interference Fit (mm) max./min.		
119			
120	O Puller Holes		
121			
122	O Balancing Holes		
123			
124			
125			
126			
127	**Coupling Guard**		
128			
129	O Supplier		
130			
131	O Flanged Cylindrical O Base Mount		
132			
133	O Air Tight		
134			
135	O Oil Tight		
136			
137	O Spark Resistant		
138			
139			
140			
141			
142			
143			

157		
158		
159		
160	**Accessories**	
161	O One set of plug and Ring gauges	
162		
163	O Drill template (for integral flanged hubs)	
164		
165	O Two-Piece Stop Rings	
166	O One Pump set	
167	(to include Hand Hydraulic Pump, Pressure Gauge,	
168	Fittings, and Hose)	
169	O Emergency Drive	
170		
171		
172	Applicable Specifications	Preparation for Shipment
173		
174		
175		O Expected Storage Time _____
176		
177		Shipping: O Dom Storage: O Indoor
178		O Exp O Outdoor
179		
180		
181		
182		
183		
184		
185		
186		

Bibliography

1   AGMA Standard 515: *Balance classification for flexible couplings*, (American Gear Manufacturers Association).

2   API Standard 617: *Special purpose couplings for refinery services*, (American Petroleum Institute).

3   AIRAPETOX, E.L., *et al.* 'Load capacity of toothed couplings', *Russian Engng J.*, 1972, **52**, 18–21.

4   ANDERSON, J.H. 'Turbocompressor drive couplings', ASME *J. Engng Power*, 1962, 115–123.

5   BARK, R. 'Schaden infolge unzureichender schmierung', *Der Maschinenschaden*, 1965, **38**, 183–184.

6   BATTERHAM, A.J. 'Turbine generator alignment using a laser-based optical system', IMechE paper C5/83.

7   BENKLER, H. *Der mechanismus der lastverteilung an bogenverzahnten zahnkupplungen*, Dissertation, Technical Hochschule Darmstadt, 1970.

8   BENKLER, H. 'Zur auslegung von bogenverzahnten zahnkupplungen', *Konstruktion*, 1972, **24**, 326–333.

9   BERG, M.S. 'Get the most out of gear tooth couplings', *Power Transmission Design*, 1961, December, 29–32.

10 BLOCH, H.P. 'How to uprate turbo equipment by optimized coupling selection', *Hydrocarbon Processing*, 1976, January, 87–90.

11 BLOCH, H.P. 'Why properly rated gears still fail', *Hydrocarbon Processing*, 1974, December, 95.

12 BORISOV, B.R. and SUNTSOV, V.N. 'Effect of couplings on critical velocity of a two-span shaft', *Russian Engng J.*, 1966, **XLVI**, 14–16.

13 BOYLAN, W. 'Marine application of dental couplings', Paper 26, Society of Naval Architects and Naval Engineers, Marine Power Plant Symposium, May 1966.

14 BOYLAN, W. 'Sliding velocity as a criterion for coupling design'. Report Code No. 2732. 9410/F 9420/F. Ft—12. Naval Boiler and Turbine Laboratory, Philadelphia, PA.

15 BREUER, K. 'Kupplungen fur walzwerksantriebe'. *DEMAG-Nachrichten*, 1959, No. 155, 20–25.

16 BREWER, 'Lubricating flexible couplings', *Diesel Power*, 1953, **31**, 56–61.

17 BS 3170: *Flexible couplings for power transmission*, (British Standards Institute).

18 BS 3853: 1966, *Mechanical balancing of marine main turbine machinery*, (British Standards Institute).

19 BROERSMA, G. 'Chapter on analysis of Bibby type couplings', in *Couplings and Bearings*, (Technic Publications H Stam Culemborg, Netherlands) 1968.

20 BROERSMA, G. 'Note on flexible couplings. Analysis of Bibby coupling', *Rev. M. Mech.*, 1972, **18**, 137–142.

21 BROERSMA, G. 'The design and manufacture of gear type couplings, Part I', Antriebstechnik; 1965, **4**, 129–138, (CEGB Translation 6486); Part II, *ibid.*, 1965, 4, 363–368, (CEGB Translation 6487); Part III, *ibid.*, 1965, 4, 453–457, (CEGB Translation 6488).

22 BROOKS, J.J., and SELLORS, P. 'An investigation into materials for use in claw couplings of turbo-generators', *Lubrication and Wear, Second Convention*, 2nd Conv., V.178., Pt. 3N., 1964, P. 45–50 *Proc. Inst Mech Engrs*, 1963–64, **178**, Pt 3N, 45–46.

23 BURSHTEIN, 'Stresses and deformations in bellows type flexible couplings' (in Russian), *Mashinostroenie*, 1964, 92–101.

24 CALISTRAT, M.M. 'Grease separation under centrifugal forces', ASME Paper Number 75—PTG—3; presented at the Joint ASLE-ASME Lubrication Conference, Miami, Florida on 21–23 October, 1975.

25 CALISTRAT, M.M. 'What causes wear in gear-type couplings', *Hydrocarbon Process-ing*, 1975, January, 53–57.
26 CALISTRAT, M.M., and LEASEBURGE, G.G. 'Torsional stiffness of interference fit connections', Paper 72—PTG—37, presented at the Mechanisms Conference and International Symposium on Gearing and Transmissions, ASME, in San Francisco, California, 8–12 October, 1972.
27 CHARTAN and WHITE. 'Flexible couplings for marine installations — testing and application', Institute of Marine Engineers, Paper 9, February 1965.
28 CHASE, T.W. 'Wie werden zahnkupplungen betriebssicher?' *Die Maschine*, 1968, **22**, 75–78.
29 CHASE, T.W., and KIMMEL, J.J. 'Maintenance of gear type couplings', *Iron and Steel Engineer*, 1964, **41**, 72–84.
30 CHEN, Y.N. 'The instability of the pump system induced by the orbital movement of the gear coupling spacer', *Pumpentagung*, Karlsruhe, Paper K6, 1973.
31 CHROBOT, B. 'The removal of large shrink-fitted couplings by hydraulic pressure', *Engineers Digest*, 1969, **30**.
32 CLAPIS, A., LAPINI, G. and ROSSINI, T. 'Diagnosis in operation of bearing mis-alignment in turbogenerators', ASME Paper 77—DeT—4.
33 CONRADT, J. 'Kinematic study of gear couplings' (in German). *Konstruktion*, 1956, **8**, 271–275.
34 CONTI-BARBARAN, D. 'Some remarks on tooth-type flexible couplings', Fourth Round-Table Discussion on Marine Reduction Gear, Schloss Brestenberg, Switzer-land, September 1961 (*The Marine Engineer and Naval Architect*, November 1963, 544–546).
35 CUMMINGS, J.M. 'Users experience with instrumented couplings for continuous movement of manpower and alignment on large turbomachinery trains', ASME Paper 80—C2/DET—52.
36 DODD, V.R. 'Shaft alignment monitoring cuts costs' *Oil and Gas Journal*, 25 Sep-tember 1972.
37 DUDLEY, D.W. 'How to design involute splines' and 'When splines need stress control', *Production Engineer*, October 1957, 75–80 and December 1957, 56–62.
38 ESSINGER, J.N. 'A new approach to hot alignment of turbomachinery'.
39 FILEPP, L. 'Flexible power transmission couplings', *Des Engng*, 1965, **61**, 125.
40 FILEP, L. 'Lubricant as a coolant in high speed gear couplings', *J. Lubric. Technol.*, 1970, **92F**, 178–179.
41 FUNK, W. 'Untersúchungen der reibkorrosion bei schrumpfund spannverbindungen', *Forschungsbericht*, Nr. 2–129/1(1966), der Forschungsvereinigung Verbrennungskraft-maschienen e.V., Frankfurt a/M.
42 GALLAS 'Accouplements elastiques avec eliments en elastomere'. *Rev. M. Mech.*, 1972, **18**, 143–154.
43 GEFIATOV, 'Calculation of the compensating capability of an elastic coupling using tangentially disposed connecting links', *Chem. Pet. Eng.*, 1972, **8**, 117–119.
44 GERSHEIMER, 'Flexible couplings, Design of elastomeric type', *Mach. Des.*, 1961, **33**, 154–159.
45 GIBBONS, 'How they are using the flexible diaphragm couplings', *Power Transmission Design*, 1969, 42–43.
46 GROSSKORDT, D. 'Neue und bewahrte kupplungen', *Ubersee-Post*, 1961, No. 53, 16–17.
47 HANISCH, F. 'Zahnradschaden und ihre beeinflussung durch schmierstoffe', *Erfah-rungsberichte*, 1965, No. 2, der Allianz Versicherungs–AG.

48 HASHEMI, Y. 'Alignment changes and their effects on the operation and integrity of large turbine generators; experience in its CEGB South Eastern Regions', IMechE paper C3/83.
49 'High speed Flexible Couplings – An Engineering Seminar', 11–13 June 1962, Pennsylvania State University (Engineering Proceedings, 1962, 41).
50 HOFFMAN, H.N. 'What makes flexible couplings flex', *Power Engineering*, 1960, **3**, 77–78.
51 HUNT, 'Constant velocity shaft couplings – a general theory', ASME Paper No. 72-Mech-10, 1972.
52 HYLER, J.E. 'Gear type flexible couplings', *Southern Power and Industry*, 1956, 52.
53 'Guide for shaft couplings for special purpose rotary machines', *ICI PLC Dept of Design Engineering*.
54 ISO 1940: *Balance quality of rotating rigid bodies* (International Standards Organization).
55 ISO 4863: *Specifying characteristics of resilient shaft couplings* (International Standards Organization).
56 JACKSON, C. 'Techniques for alignment of rotating equipment', *Hydrocarbon Processing*, Jan 1976.
57 JONCKHEERE, 'Le probleme general de l'accouplement des machins tournantes', *Rev. M. Mech.*, 1972, **18**, 129–136.
58 JONES, T.P. 'Design operating experience and development potential of main propulsion epicyclic gears', *Trans I. Mar. E.*, 1972, **84**, 476.
59 JONSSON, H., and PELTERSON, C.O. 'Oil injection method applied to flanged and gear-type couplings', *Ball Bearing Jnl*, **192**, 19–27.
60 KER WILSON, W. 'Practical Solutions of Torsional Vibration Problems', 1968, **45**, 157–169.
61 KRAEMER, K. 'Are couplings the weak link in rotating machinery systems?' *Hydrocarbon Processing*, February 1974, 97–100.
62 KRAEMER, K. 'New coupling applications or application of new coupling designs', Proc. 2nd Turbomachinery Symp., Gas Turbine Laboratories, Texas A and M University, USA.
63 KU, P.M., and VALTIERRA, M.L. 'Spline wear – effects of design and lubrication', *J. Engng Ind.*, 1975, 1257–1265.
64 LEES, A.W., and HAINES, K.A. 'Torsional vibrations of a boiler feed pump', ASME Paper No. 77-DET-28, for ASME Vibration Conference, Chicago, September 1977.
65 LEVIN Z.M. 'Load concentration along a toothed (splined) coupling', ???????????
66 Lloyds Register of Shipping. Rules and Regulations, Part 5: *Main Auxillary Machinery*.
67 McCHEYNE, R.M. 'A theoretical and experimental investigation of vibration in a boiler feed pump drive line', *Vibrations and Noise in Pump, Fan, and Compressor Installations*, 1975, IMechE, London, 11–20.
68 McMATH, R.R. 'An experimental study of fretting wear in gear tooth flexible couplings', *ASLE Trans*, 1961, **4**, 197–212.
69 Michael Neale & Associates Ltd, *Proc. of the Intl Conf. on flexible couplings for high powers and speeds*.
70 MOHLE, H. 'Assessment criteria for high speed gear type couplings' (in German), *Konstruktion*, 17, 61–66 (CEGB Translation 6489).
71 MOKED, I. 'Toothed couplings – analysis and optimization', *J. Engng Ind.*, 1968, **90**, 425–435.
72 MULLER, G. 'Bewegliche doppelzahnkupplung, insbesondere fur hohe drehzahlen', DWP 21857, KI.47c4. Maschinenbautechnik, 1963, **12**, 3.
73 NESTORIDES, E.J., *BICERA Handbook of Torsional Vibration*.
74 NIBERG, N. Ya. 'The permissible misalignment of gear and other types of coupling', *Machines and Tooling*, 1961, **32**, 10–13.

75 PADDON, F.H. 'Application and selection of gear type spindles', *Iron and Steel Engineer*, 1960, **9**, 91–100.
76 PAHL, G. 'The operating characteristics of gear-type couplings', Proc. of 7th turbo-machinery Symposium.
77 Pennsylvania State University, 'High speed flexible couplings' *Engineering Proceedings*, 1963, 41.
78 PLOTNIKOV, V.A. 'Geometrical design of an involute splined joint used in gear type couplings (in Russian), *Mashionostroenie*, 1965, No. 6, 21–31 (CEGB Translation 6493).
79 PRAY, L.H. and GOODY, E.W., 'All metallic laminated flexible element couplings', ASME Paper 72-PTG-38, 1972.
80 'Pump coupling needs careful line-up', *Power*, March 1975, 45–47.
81 RAI, K.K. 'Development, design and performance of toothed couplings', *J. Inst. Engng (India), Mech. Eng. Div.*, 1975, **53**, 244–248.
82 RAO, J.S., and CARNEGIE, W. 'Non linear vibrations in a flexible coupling', *Shipping World and Shipbuilder*, 1970, **163**, 657–659.
83 REED, F.E. and BATTER, J.F. 'An experimental study of fretting and galling in dental couplings', ASLE Gear Symposium, Chicago, January 1959.
84 RENZO, P.C. KAUFMAN, S., and DE ROCKER, D.E. 'Gear couplings', *J. Engng Industry*, 1968, **90**, 467–474.
85 ROTHFUSS, N.B. 'Design and application of flexible diaphragm couplings to industrial – marine gas turbines', ASME Paper no. 73-GT-75, 1973, 6–7.
86 ROTHFUSS, N.B. 'Design criteria and table for selecting high speed power couplings', *Prod. Eng.*, 1963, **34**, 55–66.
87 ROTHFUSS, N.B., KAISER, ?, BREENING, ?, and GIBBONS, ? 'Design and application of metallic flexures for equipment with specific life requirements', SAE Air transport and Space Meeting, April 1964, Paper 871A.
88 RUGGEN, W., and STUBNER, K. 'Zahnkupplungen – eine ubersicht uber bestehende bauarten', *Industrieblatt*, 1964, **64**, 318–324.
89 KYAKHOVSKII, O.A. 'Determining the compensating capacity of flexible disc couplings' (in Russian), *Mashinostroenie*, 1964, No. 12, 59–70 (CEGB Translation 6492).
90 SCHACH, 'Calculation of flexibility of various types of flexible couplings' (in German), *TZ fuer Praktische Metallbearbeitung*, 1966, **60**, 227–233.
91 'Schaden an einer zahnkupplung durch ankornungen', *Der Maschinenschaden*, 1970, **43**, 159–160.
92 SCHALTZ, *Kupplings-Atlas (Manual of Machine Couplings)*, 1969, 3rd Edition (in German) *AGT-Verlag Ludwigsburg*.
93 SCHENK, R. 'Zahnkupplungspatentschrift der Firma American Flexible Coupling Company, Pennsylvanien', 1801 Pittsburgh Avenue, Erie, Pennsylvania, 1966.
94 SEIREG, 'Static and dynamic behaviour of flexible torsional couplings with non-linear characteristics', ASME Paper no. 59-SA-8, 1959.
95 SHIRAKI, K. and UMERBURA, S. 'On the vibration of a two rotor system connected by a gear coupling', *Mitsubishi Technical Review*, 1970, **1**, 22–23.
96 Sier Bath Gear Company, 'For longer lived couplings, a new curve on gear teeth', *Product Engineering*, 25 April, 1966, 55.
97 STAEDELI, O. 'Tooth couplings', Maag Gear Xo., Zurich, October 1973.
98 STAKHANOV, I.E. 'A geared coupling with an elliptical tooth form and a pneumatic seal', *Russian Engineering Journal*, 1962, **8**, 8–11.
99 STOCKIGHT, A. 'Zahnkupplungs-Patent No. BRD-AS 1283 607, KI. 47c, Gr.3–18 (1968)', *Maschinenbautechnik*, 1970, **19**, 218.

100 STOLZLE, K. 'Flexible couplings', *Technische Mitteilungen*, 1962, **7**, 316–325 (BISI Translation 7173).
101 Texas Oil Corporation 'Flexible couplings I' *Lubrication*, 1962, **17**, 29–40; 'Flexible couplings II', *Lubrication*, 1962, **17**, 41–48.
102 VANCE, J.M., 'Influence of coupling properties on the dynamics of high speed power transmission shafts', ASME Paper no. 72-PTG-38, 1972.
103 VEITZ, 'Dynamics of machine assemblies with non-linear couplings', IMechEng., V.6, No. 4, Winter 1971, 367–382.
104 WATTNER, K.W. 'Failure analysis of high speed couplings', *Hydrocarbon Processing*, January 1976, 77–80.
105 WEATHERFORD, W.D., Jr, VALTIERRA, M.L. and KU, P.M. 'Experimental study of spline wear and lubrication effects', *ASLE Trans*, 1966, **9**, 171–178.
106 WEBB, S.G. and CALISTRAT, M.M. 'Sludge accumulation in continuously lubricated couplings', Paper to the 27th Annual Petroleum Mechanical Engineering Conference, ASME, New Orleans, September 1972, 17–21.
107 WILDHAGEN, 'Membrankupplungen', *Antriebstechnik*, 1969, **8**, 228–229.
108 WOLFF, P.H.W. 'The design of flexible disc misalignment couplings', *Proc. Instn Mech. Engrs*, 1951, **165**, 165–175.
109 WOODCOCK, J.S. 'Balancing criteria for high speed rotors with flexible couplings', *Vibrations and Noise in Pump, Fan, and Compressor Installations*, 1975, IMechE, London, 107–114.
110 WRIGHT, J. 'Which flexible coupling?', *1971/1972 Power Transmission and Bearings Handbook*, Industrial Publishing Co., Division of Pittway Corporation.
111 WRIGHT, J. 'Which shaft coupling is best – lubricated or non-lubricated?', *Hydrocarbon Processing*, April 1975, 191–193.
112 YAMPOL'SKII, I.E. *et al*. 'Formation of disturbing forces in toothed clutches', *Russian Engineering Journal*, 1960, **LV**, 13–17.
113 ZURN, F.W. 'Crowned tooth gear type couplings – development and application for the steel industry', *Iron and Steel Engineer*, August 1957, 98–116.